Introduction to DNA Self-Assembled Computer Design

For a list of recent related Artech House titles turn to the back of this book.

Introduction to DNA Self-Assembled Computer Design

Christopher Dwyer
Alvin Lebeck

ARTECH HOUSE
BOSTON | LONDON
artechhouse.com

Library of Congress Cataloging-in-Publication Data
A catalog record for this book is available from the U.S. Library of Congress.

British Library Cataloguing in Publication Data
A catalogue record for this book is available from the British Library.

Cover design by Yekaterina Ratner

ISBN 13: 978-1-59693-168-8

© 2008 ARTECH HOUSE, INC.
685 Canton Street
Norwood, MA 02062

All rights reserved. Printed and bound in the United States of America. No part of this book may be reproduced or utilized in any form or by any means, electronic or mechanical, including photocopying, recording, or by any information storage and retrieval system, without permission in writing from the publisher.

All terms mentioned in this book that are known to be trademarks or service marks have been appropriately capitalized. Artech House cannot attest to the accuracy of this information. Use of a term in this book should not be regarded as affecting the validity of any trademark or service mark.

10 9 8 7 6 5 4 3 2 1

To Andrea and Ian
To Mitali, Niel, and Kiron

Portions of this manuscript are based on works copyrighted by IEEE, ACM, and the Institute of Physics

© 2005, IEEE. Reprinted, with permission, from Self-Assembled Architectures and the Temporal Aspects of Computing, Chris Dwyer, Alvin R. Lebeck, and Daniel J. Sorin, in *IEEE Computer*, 38 (1), pages 56–64, January 2005.

© ACM, 2006, NANA: A nano-scale active network architecture, in *ACM Journal on Emerging Technologies in Computing (JETC)*, {Vol 1, Issue 1, January}, http://doi.acm.org/10.1145/1126257.1126258.

© ACM, 2007, A Self-Organizing Defect Tolerant SIMD Architecture, in *ACM Journal on Emerging Technologies in Computing (JETC)*, {Vol 3, Issue 2, July}, http://doi.acm.org/10.1145/1265949.1265956.

© Institute of Physics, 2007, Scalable, Low-cost, Hierarchical Assembly of Programmable DNA Nanostructures, C. Pistol and C. Dwyer, *Nanotechnology*, vol. 18, 125305–9, 2007.

© 2006, IEEE. Reprinted, with permission, from Design Automation for DNA Self-Assembled Nanostructures, C. Pistol, C. Dwyer, A. R. Lebeck, in *Proceedings of the 43rd Design Automation Conference (DAC)*, July, 2006.

© 2006, IEEE. Reprinted, with permission, from Self-Assembled Networks: Control vs. Complexity, Jaidev Patwardhan, Chris Dwyer, Alvin R. Lebeck. *1st International Conference on Nano-Networks (NANONETS)*, September 2006.

© 2006, IEEE. Reprinted, with permission, from Design and Evaluation of Fail-Stop Self-Assembled Nanoscale Processing Elements, J. Patwardhan, C. Dwyer, A. R. Lebeck, in *IEEE International Workshop on Design and Test of Defect-Tolerant Nanoscale Architectures (NANOARCH '06)*, June 2006.

Contents

	Acknowledgments	*xiii*
1	**Introduction**	**1**
1.1	Top-down vs. Bottom-up	2
1.2	Review of Conventional Fabrication Techniques	3
2	**DNA Self-Assembly**	**7**
2.1	Nucleotides, Oligos, and the Double Helix	9
2.1.1	Thermodynamics	9
2.1.2	Sequence Design	10
2.2	DNA Motifs	10
2.2.1	Related DNA Nanostructures	11
2.3	Metrics and Design Rules	11
2.3.1	Metrics	12
2.3.2	Design Rules	13
2.3.3	Target Nanostructure	14
2.4	Design Automation Methods	14
2.4.1	A Thermodynamic Optimization Tool	15
2.4.2	Alternative Designs	16
2.4.3	Design Evaluation	18
2.5	Hierarchical Assembly	23
2.5.1	Generic Linkers	25
2.5.2	Fully Specific Linkers	27
2.6	Summary	29
	References	29

3	**Self-Assembled Circuitry and Design**	**33**
3.1	Introduction	33
3.2	Related Work	33
3.2.1	Mesoscale Self-Assembly	33
3.2.2	Nanoparticle-DNA Conjugates	34
3.3	DNA Scaffolding Structures	34
3.4	Implications for Nanoelectronic Circuit Architecture	35
3.4.1	Regularity	35
3.4.2	Complexity	35
3.4.3	Defect Tolerance	36
3.5	Nanoelectronic Circuit Building Blocks	37
3.5.1	Exploiting Regularity: A Replicated Unit Cell	38
3.5.2	Introducing Complexity: An Aperiodic Pattern for Interconnection Cells	39
3.6	Large-Scale Interconnection of Circuit Nodes	39
3.7	DNA Design Flow	40
3.7.1	Overview	41
3.7.2	Circuit Design	41
3.7.3	Device Design	43
3.7.4	Self-Assembled DNA Design	43
3.8	Case Studies	45
3.9	Conclusions	49
References		49
4	**Architectural Implications of Self-Assembly**	**51**
4.1	Technology Implications	52
4.1.1	Small-Scale Control	52
4.1.2	Large-Scale Randomness	53
4.1.3	High Defect Rates	53
4.2	Architectural Challenges	54
4.3	Opportunities for New Architectures	55
4.3.1	The Temporal Aspects of Computing	56
4.3.2	Extending the Temporal Structure	59

4.4	Summary	60
References		60

5   Oracles and At-Fabrication Computation   63

5.1	Introduction	63
5.2	System Overview	63
5.3	A Simple Oracle	64
5.4	Implementation	65
5.5	At-Fabrication Computation Requirements	67
5.6	Generalization of the Oracle	68
5.7	Hamiltonian Path Oracle	69
5.8	Block Edit Oracle	70
5.9	Purely Postfabrication Computation	72
5.10	At-Fabrication Components of Optimal Block Edit Solutions	73
5.10.1	At-Fabrication Block Partitioning	73
5.10.2	At-Fabrication Block Reordering	74
5.10.3	Alternative Points in the Temporal Design Space	75
5.11	Summary	77
References		77

6   The Distributed Array Multiprocessor   79

6.1	Introduction	79
6.2	System Overview	79
6.3	Execution Model	80
6.4	Hardware Design	83
6.4.1	Modular Assembly	83
6.4.2	Monolithic Assembly	85
6.4.3	Output Methods	88
6.5	System Operation	90
6.5.1	A DAMP Instruction Set	95
6.6	Performance	98
6.6.1	Simple Comparisons	100

6.6.2	Blind Decryption of the Data Encryption Standard	102
6.7	Summary	103
	References	104

7 A Nanoscale Active Network Architecture — 107

7.1	Introduction	107
7.2	A General-Purpose Architecture for Self-Assembled Nano-Electronics	107
7.3	System Model	108
7.4	Execution Model	110
7.5	Instruction Set and Packet Formats	111
7.6	Interconnection Network: Finding Resources for Execution	114
7.6.1	Imposing Structure with Gradients	115
7.6.2	Execution Packet Routing	118
7.7	Memory	119
7.7.1	Memory Allocation	119
7.7.2	Interfacing Execution and Memory	120
7.7.3	Routing Memory Packets	121
7.8	Packet Instantiation and Chaining	121
7.9	Improving Node Utilization	122
7.10	Preliminary Evaluation	122
7.10.1	Node Floorplan	123
7.10.2	Simulation Framework	124
7.10.3	Fibonacci	124
7.10.4	String Match	126
7.11	Discussion	128
7.12	Conclusions	129
	References	129

8 A Self-Organizing Defect Tolerant SIMD Architecture — 131

8.1	DNA-Based Self-Assembled Nanoscale Systems and the Architectural Implications	132

8.2	System Overview	135
8.3	Node Microarchitecture	136
8.3.1	Data Path	136
8.3.2	Control	137
8.3.3	Internode Communication	138
8.3.4	Circuit Size and Power Estimates	139
8.3.5	Summary	140
8.4	System Configuration	140
8.4.1	Logical Structures and Defect Isolation	140
8.4.2	Configuring Processing Elements	141
8.5	System Architecture	143
8.5.1	Instruction Set Architecture	143
8.5.2	Execution Model	144
8.5.3	Instruction Execution Example	145
8.5.4	Evaluation	146
8.5.5	Methodology	146
8.5.6	Results	148
8.5.7	Sensitivity Analysis	153
8.5.8	Defect Tolerance	160
8.5.9	Result Summary	161
8.6	Limitations and Future Work	162
8.7	Related Work	163
8.8	Conclusions	163
8.9	Programming SOSA—Matrix Multiplication	164
References		167

9 Overcoming Randomness with Increased Complexity — 171

9.1	Introduction	171
9.2	Design and Evaluation of Fail-Stop Nodes	172
9.2.1	Defect Isolation Using Reverse Path Forwarding	173
9.2.2	Fail-Stop Nodes	174
9.2.3	Evaluation	181
9.2.4	Related Work	186
9.2.5	Summary	187

9.3	Self-Assembled Networks: Control Versus Complexity	187
9.3.1	Node and System Architecture	188
9.3.2	Networks of Self-Assembled Processing Elements	188
9.3.3	Experimental Setup and Evaluation	191
9.3.4	Summary	195
References		195

Appendix: Laboratory Methods in DNA Self-Assembly — **197**

A.1	General Lab Practices	197
A.2	Creation and Storage of Buffers	199
A.3	Aliquotting DNA Strands	200
A.4	Tile Annealing	202
A.5	Grid Annealing	203

About the Authors — **209**

Index — **211**

Acknowledgments

We thank our various collaborators over the years for their contributions that led to the body of work described in this book. We specifically thank Jaidev Patwardhan, Dan Sorin, Constantin Pistol, and Vincent Mao for their contributions.

The research reported is supported by NSF grant CCR-0326157, NSF grant CCF-0702434, the Duke University Provost's Common Fund, AFRL contract FA8750-05-2-0018, Microsoft, Agilent, and equipment donations from IBM and Intel.

1

Introduction

An Introduction to DNA Self-Assembled Computer Design describes how biological molecules and nanotechnology can impact the ways we design, build, and use computer systems. From macroscopic to molecular scales, self-assembly can be found creating the complex structures and functions that underpin our world. Self-assembly, however, is a process that receives little coverage in traditional engineering yet has impact on almost every engineered system. Biology is ripe with examples of complex self-assembly and these examples inspire us to approach the engineering of complex computer systems in new ways. Written for a general technical audience, this book is intended for readers who wish to learn what they need to understand this fast growing and ground breaking field.

Readers who may have forgotten some of their introductory biology and chemistry will find some help in Chapter 2, which is a survey of the self-assembly process that we build upon in later chapters. Readers who are unfamiliar with techniques for integrated circuit design will benefit from a solid review of microelectronics and VLSI design. To augment such a review we provide a brief description of conventional circuit fabrication techniques at the end of this chapter.

Device manufacturing at dimensions that approach tens of nanometers has significant challenges that stem from the finite size and structure of matter. In spite of such challenges, commercial microprocessor manufacturers continue to achieve ever smaller device feature sizes but at the cost of escalating manufacturing complexity. Thus, new methods for building computers that can reduce manufacturing costs and achieve performance equal to or greater than conventional systems will create new opportunities for computing. We will highlight some of these new opportunities and discuss how self-assembly can enable new modes of computation in Chapter 7. First, we begin our discussion of self-assembly by contrasting conventional top-down fabrication methods against the bottom-up techniques that will play a role in self-assembled computer fabrication.

1.1 Top-Down Versus Bottom-Up Fabrication

Top-down fabrication methods impose control over the placement, composition, and structure of materials from macroscopic bulk stock. Subtractive or additive processes, such as those employed in photolithography, are fundamentally top-down because structure is created by selectively depositing or removing bulk solids (or liquids) to form desired patterns. Imprint lithography is also a top-down process since it uses a macroscopic pattern (i.e., a stamp or master) to impress structure upon a uniform (bulk) surface. The common feature of top-down processes is a continuous reduction in the characteristic length scale of material structure from the macroscale to the molecular scale. For example, the imprint master must mechanically interface with the macroscopic stamp aligner and, by a smooth transition of size scales from the bulk stamp to the nanoscale pattern on the underside of the stam, come into commensurate contact with the target surface. For this process to be useful, the features of the stamp must convey nanoscale structural properties to the surface with high fidelity (e.g., the target surface must take on the same pitch, aspect ratio, etc. as the features on the stamp pattern.) The inherent challenge with top-down fabrication is that as feature sizes approach molecular scales (i.e., 0.1 to 10 nm) materials no longer behave in the same rational and deterministic fashion that we encounter at macroscopic dimensions. Thus, a smooth transition from the macroscale to the molecular scale will require innovation to maintain the traditional device scaling (e.g., from 45 to 32 nm) that has driven the computer manufacturing industry.

The alternative to top-down fabrication is bottom-up fabrication whereby structure is created by the assembly of precursors (or blocks) that grow outward from a nucleation site. The everyday variety of bottom-up fabrication uses precursors built from top-down methods. For example, the stones in a stone wall are laid in place such that the wall grows outward from its foundation. Contrast this method with a top-down method where a wooden form is built to define the wall. In fact, form-built structures rely on both top-down fabrication (e.g., the form) and bottom-up fabrication through the molecular adhesion of the fill material (e.g., concrete). Unlike top-down fabrication, bottom-up processes typically rely on local interactions between precursors (and their environment) alone to create large-scale structure.

Self-assembly is a mechanism for bottom-up fabrication in which the precursors are only weakly controlled by external environmental parameters such as temperature or reaction time. A comprehensive survey of bottom-up self-assembly is beyond the scope of this introduction but can be found in any good text on chemical synthesis. Fundamentally, thermodynamics governs (and drives) precursors to form structure and, by careful design, can yield useful products.

The chapters in this book are organized to build from the low-level technological details of self-assembly and transition to circuit design, and to finally develop models of the computer architectures that are uniquely enabled by self-assembly. In Chapter 2 we discuss deoxyribonucleic acid (DNA) self-assembly as a means to form complex molecular structures and we provide several laboratory protocols for these in the Appendix. We cover the aspects of conventional circuit design that change within the context of self-assembly in Chapter 3, and in Chapter 4 we identify several technological challenges to building self-assembled computer systems. We present a survey of several proof-of-concept architectures that resolve many of these challenges in Chapters 5 through 8. Chapter 9 concludes the book with a study of how the inherent randomness in self-assembly can be mitigated by defect tolerance and a limited form of control over the self-assembly process.

1.2 Review of Conventional Fabrication Techniques

The dominant technique in use today to create integrated circuits is photolithography. There are many varieties of this technique and each employ photons to optically induce chemical changes in a polymer surface. The polymer surface, also known as a resist layer, serves to selectively protect regions of the underlying surface (or wafer) from subsequent chemical processing. Resist layers are typically deposited by a process known as spin coating whereby a reliable and repeatable layer thickness can be achieved by controlling the rotational velocity profile in time of the wafer after a controlled volume of resist (in a liquid state) has been poured onto the wafer. The entire wafer (with resist layer) is then treated to render the layer sensitive to light. The wafer is then exposed to an optical light field (either directly, through reduction optics, or by optical interference) that is spatially modulated to create a desired pattern. The pattern determines where subsequent processing will alter the underlying wafer surface. Once exposed, the resist layer will either be susceptible to removal by a strong solvent (e.g., positive resists) or not (e.g., negative resists), during a process called development. Many resist layers require processing after exposure (but before development) to produce deep profiles.

Once the resist layer has been processed to selectively protect the wafer, materials such as metals, etchants, dielectrics, or dopants are deposited (or reacted) with the exposed surface. These materials render the surface chemically distinct only in areas where the resist was not protecting the wafer. Each pattern and material is chosen to create the precise structure on the wafer required to realize the final integrated circuit.

After material processing the wafer undergoes a cleaning process that removes the resist layer and prepares the underlying surface for another round of patterning. These rounds continue for as many layers as are required in the design. Today's microprocessors require more than 40 such rounds to yield a final product. An important process called mask alignment is performed between rounds to ensure that subsequent patterns are precisely aligned with previously generated structures on the wafer. This process cannot be perfect due to finite mechanical tolerances in the equipment and must be mitigated by engineering the designed patterns to account for the misalignment. For example, if layer-to-layer alignment at a given foundry is limited to ±20 nm, designers must take care to ensure that structures that vertically span multiple layers be at least 400 nm^2 square (per layer spanned) to guarantee overlap between the two layers.

Imprint lithography differs from photolithography only in the way that the resist layer is patterned. Under high pressure, an imprint master (or stamp) is forced into a heated resist layer. The resist is designed to be viscous at this temperature and under pressure it will be displaced by the master. The places where the master and wafer come into contact will exclude the resist and, after cooling, the resist will retain the shape of the master. It is not a trivial process to remove the master without disturbing the resist but methods exist to facilitate this step. The ultimate resolution of patterned features with imprint lithography is better than those achieved by photolithography but reduction optics cannot be used. Photolithography is limited to a fraction (~$\frac{1}{10}$) of the wavelength of the exposing light. Theoretically, imprint lithography is limited only by the ability of the master to displace the resist and thus, can generate molecular-scale patterns. However, in practice, the imprint master must be generated by some other method which typically has a worse resolution limit. Today, most masters for imprint lithography are made by electron beam lithography and have a short lifetime due to contact with the resist layer.

Electron beam lithography (EBL) is similar to photolithography but differs in the exposure process. A focused electron beam is used to raster a pattern onto the resist layer and the interaction of the beam current passing into the resist layer (at a given dose) alters the resist to make it susceptible (or not) to development. EBL can achieve higher pattern resolution than photolithography because the wavelength of the electron is smaller than the wavelength of the photons used in standard photolithography. Research into extreme ultraviolet and soft x-ray exposure (with wavelengths comparable to or better than an electron beam) has demonstrated proof-of-concept systems but these have not yet seen wide adoption.

Regardless of the method used to create the circuits on the wafer a final passivation layer is deposited and patterned to protect the devices from external contamination and damage. The devices (and systems) on the wafer are then tested and working chips are cut from the wafer and packaged.

The ease with which silicon oxide can be grown, metal contacts deposited, and dopants implanted have made silicon the dominant material system for computer fabrication. Complementary metal oxide semiconductor (CMOS) devices are readily manufactured on silicon wafers and are the most common devices used today. Micro-scale CMOS devices are excellent for computing because of low static and dynamic power consumption, high switching speeds, and high density (i.e., each device can be small). However, as CMOS device sizes approach the molecular scale, many of these advantages deteriorate. Further, new failure modes such as gate dielectric breakdown begin to reduce the reliability of devices and curtail the longevity of CMOS systems.

Much of what we call conventional fabrication technology uses a top-down approach. However, the chemical processes that form the devices and interconnect are inherently bottom-up. The merger between the two is achieved by patterns in resist layers similar to the wooden forms used to confine concrete from our earlier example. Beyond this, our goal is to achieve the complexity of structure that we need for our systems but at a scale where top-down approaches fail (i.e., molecular scales). Thus, we must use bottom-up methods to build complex systems from molecular components.

2
DNA Self-Assembly

In the traditional approach to designing and fabricating computer systems, designers use a top-down scheme to specify exactly where every component should be placed and then the manufacturer fabricates the system according to those specifications. This intuitive top-down approach conforms with how we expect to design systems and it is how all current commercial computer systems have been developed. However, the semiconductor industry has identified the difficulty of continuing to shrink the feature sizes for mass-produced electronic components, and this impending roadblock has spurred research in bottom-up self-assembly approaches.

In bottom-up self-assembly the designer specifies the components but cannot stipulate exactly where each one will be placed. Rather, the components must self-assemble to form a system. The simplest form of self-assembly is random; however, this approach is limited in what it can create. Since no order is imposed on the fabrication the designer can introduce little complexity.

An alternative to random self-assembly is programmable self-assembly in which the designer specifies how the components attach to one another but not where any particular component will be placed. With programmable self-assembly the designer has some control over the fabricated system and can thus create more sophisticated structures.

One approach to achieve this programmability is to use DNA. The precise binding rules of DNA enable the creation of nanostructures with feature sizes on the order of a few nanometers. These nanostructures can then be used to organize and interconnect nanoscale components (e.g., crossed carbon nanotube FETs, ring-gated FETs, and nanowires).

The challenge in creating DNA nanostructures is to specify the appropriate DNA sequences such that the desired structure (geometry) is formed and is thermodynamically stable. To meet this challenge DNA self-assembly can exploit the common technique of hierarchically composing a small set of relatively simple motifs to create more sophisticated structures. Many parts of this design process can benefit from computer-aided design automation. However, in this chapter we focus on the key aspect of designing the DNA sequences that control how motifs can bind with each other. Specifically, we seek to find DNA sequences that minimize the strength of unintentional interactions with other motifs in the set while maximizing the strength of intentional interactions.

This chapter first presents an approach to evaluate the sequence design space to create a fixed size 60 × 60-nm grid with 20-nm pitch. This structure is composed through a hierarchical assembly of motifs (Figure 2.1). Ultimately, the final assembly step combines 16 motifs (arranged 4 × 4) to form the grid structure. For this structure we must determine the best 96 sequences according to structural and stability metrics. This can be accomplished by an optimization algorithm that is aware of both intentional and unintentional interactions and exploits parallelism to rapidly evaluate the large sequence space. These 16 motifs can be combined to form larger structures and are the basis for a cost-effective and scalable approach to patterned nanostructure creation. Figure 2.2 is a schematic of the complete hierarchical assembly starting with single strands of DNA to form motifs (i.e., cruciforms), and then 4 × 4 grids, which are then assembled into larger structures.

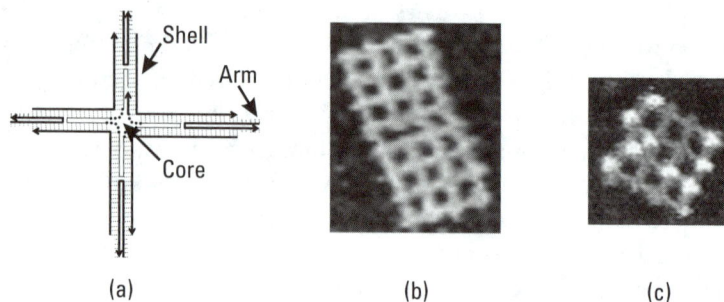

Figure 2.1 (a) Schematic of a cruciform composite motif. (b) AFM image of a hierarchical 8 × 4 grid. (c) A protein-patterned 4 × 4 grid. Each cavity in (a) and (b) is ~20 nm wide.

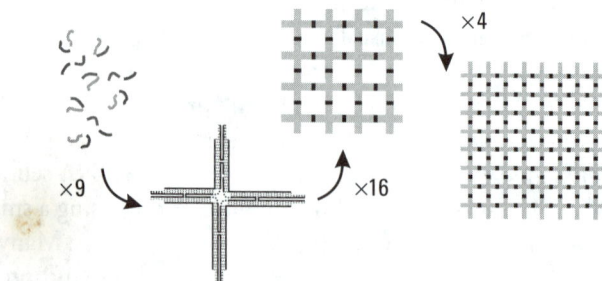

Figure 2.2 The hierarchical approach to build motifs from individual strands and then into 4 × 4 or 8 × 8 grids.

2.1 Nucleotides, Oligos, and the Double Helix

DNA is widely studied in the context of molecular genetics. However, for the fabrication of nanostructures the primary concern is how DNA can be employed as a programmable substrate and geometric tool. The following is a brief review of DNA and its properties.

DNA's basic building blocks—called nucleotides (nt)—are composed of a phosphate covalently bound to a nucleoside. Nucleosides are derivatives of a deoxyribose sugar and either a purine or pyrimidine nucleobase. The common purine nucleobases are adenine (A) and guanine (G), and the common pyrimidine nucleobases are thymine (T) and cytosine (C). A polymer of nucleotides, joined by phosphodiester bonds, is the so-called single-stranded DNA molecule. Single strands can wrap around each other to form a helical structure, or double stranded molecule, called a double helix.

The double stranded DNA structure is stable when the pairwise nucleobase interactions are "complementary"—that is, if A pairs with T and G pairs with C. Under these conditions (in the "B-form" helix) each Watson-Crick base pair (bp) is approximately 2 nm wide (diameter of the helix) and on average 0.34 nm away from adjacent bases. The helical twist of the two strands is such that a full turn occurs between every 10th and 11th base. The stability of this interaction is only approximately linear per base and depends on neighboring mismatched or complementary interactions [1]. The stability and exact dimensions, orientation, and form of the nucleobase interactions depend on several factors including the pH of the solution and the microenvironment of the DNA.

2.1.1 Thermodynamics

The central theme in the use of self-assembly for nanoscale fabrication is the application of external control over an otherwise spontaneous reaction to direct its outcome [2]. This control directs the assembly of materials into structures that are interesting and relevant to a target design problem. In the context of computer system fabrication the self-assembly is used to direct the formation of devices (e.g., transistors and wires) to create logic circuitry, memory, and I/O interfaces. Ultimately, the self-assembled DNA structure is useful only to organize the devices after which it can be rendered inert (e.g., by deposition onto a surface). The temperature of the reaction volume (i.e., the solution) is a simple control in DNA self-assembly that controls the formation of double helices from single strands as the solution temperature is cooled from a high temperature. The melting temperature (T_m) of a DNA double strand is the temperature at which the transition from single strands to double strands has reached 50%. That is, half of the single strands in solution are bound to their complementary strand when the solution temperature is exactly the melting temperature of the strand. The T_m of two strands is dependent on their sequences and the degree to which they are complementary. This simple picture is complicated by the introduction of multiple unique sequences in solution.

2.1.2 Sequence Design

A strand of DNA obeys certain thermodynamic behavior, the most important of which is that double strands form at temperatures below the T_m of the constituent single strands; this interaction can be complex if multiple unique (sequences of) DNA strands are in solution. Specification of the strand sequences provides external control over the self-assembly process (through temperature control) and determines the formation of structures (through complementarity). Good sequence design leads to a minimization of sequence mismatches, or unintentional interactions between strands of similar but not perfectly complementary sequences, at a given temperature and therefore produces a higher yield of the target structure.

Sequence design is important because it determines many aspects of the target DNA nanostructure (e.g., geometry and stability). Therefore, it is critical to have good methods for choosing sequences.

2.2 DNA Motifs

Complex designs are often created using a relatively small set of common building blocks—called motifs. DNA self-assembly can exploit this same design principle to hierarchically create more sophisticated aperiodic structures. There are many possible DNA motifs and the focus here is on only a few in the context of the target nanostructure (see Section 5). Motifs include junctions that enable three or more double stranded helices of DNA to interact and thus form specific structures (e.g., a triangle, a corner, and so on). Another important motif is a single strand of DNA protruding from a double stranded helix—called a sticky-end.

Two motifs with complementary sequences on their sticky-ends will bind to form a composite motif. Composite motifs may also have embedded sticky-end motifs and thus can also bind with other composite motifs to form another, larger, composite motif. This results in a hierarchical structure for motifs.

The cruciform motif is composed from three smaller motifs: a core, four shells, and four arms (each arm contains two 5-nt sticky-ends). Figure 2.2 shows: (a) a schematic of the cruciform motif; (b) an AFM image of a hierarchical 8 × 4 grid; and (c) a protein-patterned nanostructure, which is described in [4–6] and in this book's appendix. Although motifs provide an easy abstraction for reasoning about DNA nanostructures there are many potential issues related to sequence dependent physical (and structural) properties. For example, the above cruciform motif has a slight curvature in three-space; thus, composite motifs formed with this motif must account for this curvature to ensure the desired final geometry. Furthermore, the structural properties of any given DNA sequence can create strain in the final structure that will prohibit proper formation.

2.2.1 Related DNA Nanostructures

DNA and RNA have gained popularity as a material system for creating complex, aperiodic nanostructures due to the ease with which these materials can be synthesized and controlled [4, 5, 7–10]. The pioneering development of the DNA crossover junction enabled the rational design and synthesis of structurally rigid molecular complexes from DNA [11–16]. Such methods rely on the programmability of oligonucleotide interactions and leverage the control that complementary nucleotide sequences exert over the thermodynamics of the assembly process. Recent advances in this field have produced many examples of periodic planar DNA lattice [15, 17–21]. However, to form aperiodic 2D structures, these methods require the number of unique DNA sequences (and therefore cost) to scale with the area of the structure or the development of algorithmic self-assembly [22, 23]. To overcome such limitations a low-cost hierarchical method to fabricate large molecular weight, aperiodic structures by DNA self-assembly must be employed [6].

There are many design considerations that must be accounted for in creating a DNA nanostructure. When combined with the vast sequence space afforded to DNA these considerations motivate the need for computer aided design methods and access to large-scale computing facilities.

2.3 Metrics and Design Rules

The number of possible nanostructures that can be fabricated by motif-based DNA self-assembly is large. However, not all nanostructures can be synthesized efficiently due to the geometric and thermodynamic limitations of DNA hybridization. Such limitations require a set of experimentally verified design rules to act as templates for new designs. This section describes metrics and design rules used to design and evaluate new nanostructures.

2.3.1 Metrics

Metrics enable meaningful comparisons between designs and when coupled with baseline or reference designs they can be used to predict fabrication yields. Experimental data indicates that these metrics are correlated with high yields. The two metrics used to evaluate a design are: (1) the average single-interaction energy measure (SEM), and (2) the target-interaction likelihood measure (TLM).

SEM. The average single-interaction energy measure is an estimate of the thermal stability of the motif interactions in a design. The SEM can be used as a relative measure of stability in terms of temperature. For example, a design with a large SEM will be stable at higher temperatures than a design with a low SEM. The SEM is calculated by averaging the melting temperature of each interaction between the motifs in a design. Therefore, a large SEM indicates that the average interaction strength between motifs is also large.

While the thermal stability of a structure is important it is clear that if the structure forms incorrectly, yet with high stability, the fabrication process will produce a large fraction of flawed or defective structures. Therefore, a measure of how likely a structure will form must be coupled with the SEM to obtain a more complete characterization of the assembly process.

TLM. The target-interaction likelihood measure is an estimate of the potential for a design to assemble a correct structure. The larger this value the more likely it is that a design will form correctly. This metric is calculated as the average distance from the diagonal on a nonspecific versus specific melting temperature plot for each motif against all other motifs. Motifs that are close to the diagonal (i.e., motifs with strong nonspecific interactions) should be avoided since they will likely contribute to the formation of defective structures. This metric will always be positive.

Thus, the SEM and TLM are metrics that enable a consistent thermodynamic framework in which to compare candidate designs. They do not, however, provide insight into the geometric or structural quality of a design. The complexity of this problem motivates the use of design rules in choosing the structure and assembly order for a design. The design rules provide a template for the creation of structures from motifs and can guarantee geometric properties when the design is fabricated.

2.3.2 Design Rules

Two design rules enable the fabrication of complex nanostructures. The first is the use of a *corrugation* scheme that alternates the direction of each motif's normal vector across the structure. The second design rule describes a *thermal ordering* of motif interactions based on melting temperatures and the desired structure.

Corrugation. The term "corrugation" was first introduced by Liu et al. [24] to describe a method to combat a curling effect observed in large periodic nanostructures. Each motif has an intrinsic curvature that is sequence dependent and may or may not be approximately zero. Further, the curvature can generate a strained structure if a curved motif is forced (by adjacent motifs) into a planar shape.

A sufficiently large accumulation of strain in one direction on the structure can curl the structure into a tubule. To avoid this strain, the corrugation design rule specifies that motifs must be placed into the structure with alternating normal vectors such that their sum over the entire structure is minimal. Regardless of the curvature (or lack thereof) in the motif, the accumulation of strain can be minimized by symmetrically arranging motifs in this way. For example, consider a linear array of identical motifs each with some positive curvature. The structure will curl on itself if the motifs are aligned with each other. If the motif alignment alternates, however, they will form a straight line.

This rule can be generalized for 2D planar structures and can be applied as a template for new designs. The remaining design task (Section 2.4) is to render the generic template into real nucleotide sequences that can be used to fabricate the target structure. While the corrugation design rule will ensure that the resulting structure is planar, empirical evidence suggests that the order in which motifs assemble into the final structure plays a significant role in the defect rate of the fabrication process. This leads to a second design rule that specifies the motif assembly order.

Thermal Ordering. The DNA of each motif has a specific melting temperature below which an interaction (with other motifs) can take place and will be stable. These melting temperatures (estimated by the SEM) can be ordered from high to low and used as a criterion for picking a given sticky-end sequence and motif. Since the assembly temperature can be physically constrained to be monotonically decreasing during an experiment, the order in which motifs will assemble can be controlled by choosing sequences with descending SEMs.

Similar to the corrugation design rule, the thermal ordering design rule provides a template (in this case an assembly order) for the nanostructure and must ultimately be rendered into real nucleotide sequences. The next section describes a target nanostructure and how the metrics and design rules are applied.

2.3.3 Target Nanostructure

As a design example consider a planar grid of motifs like the ones shown in Figure 2.1. The cruciform motif described in Section 2.2 is the basic element for the grid and the design must completely specify the nucleotide sequence for each motif. The hierarchical strategy can reduce the complexity of this problem by using known nucleotide sequences for the cruciform motif and focus on the sticky-end sequences.

A 4 × 4 grid has 16 cruciform motifs and each cruciform requires four pairs of sticky-end sequences (one pair for each arm per motif). Since the motifs only bind on the interior of the grid, a total of 96 arm sequences are required (96-arm). Prior work has been limited to periodic structures in which as many as four or five motifs polymerize into 2D arrays; such systems require fewer arm sequences due to the periodic reuse of motifs [18]. A classic example of this is the "AB" system that requires 20 distinct arm sequences per an A-type and B-type motif (20-arm). These sequences must ensure that each motif will bind in only 1 of 16 positions in the grid. The corrugation and thermal ordering design rules can be used as templates for the grid at the outset and use the SEM and TLM estimates to choose from all candidate arm sequences. To maintain an optimal solution, all possible 5-nucleotide (5-nt) sticky-end sequences must be evaluated. The size of this space is exponential in the number of bases per sticky-end (i.e., large) and requires automated tools.

2.4 Design Automation Methods

Given the importance of sequence design for self-assembled systems there are a variety of tools available for this purpose [25–29]. However, these tools use heuristics, simplified interaction models (e.g., sequence text distance) or no hierarchies, which make them unable to design large systems (i.e., systems that require non-interacting sequences with greater than 1,200 base pairs). Even for small problems theses solutions do not generate a sufficient set of sticky-ends. For example, the 96-arm motif structure is too large for any of these methods.

An alternative, but trivial, method is to randomly select sequences. Using a random sequence generator thermodynamic analysis can be used to evaluate the design. The computational effort is low in this case but there are obviously no guarantees on the optimality of the resulting design.

2.4.1 A Thermodynamic Optimization Tool

New optimization tools exist to overcome the limitations of text distance and random sequence generation. Such tools use parallel implementations of an exhaustive thermodynamic search to optimize a target design against both the SEM and TLM estimators described in Section 2.3.1. The outcome is a set of sticky-end sequences that can be used to generate a set of motifs that are optimized to reduce nonspecific interactions during the assembly process. The algorithm evaluates each possible arm sequence against all other candidate sequences and motif sequences and maps their mutual interaction. Self-binding and region mismatching of up to 6 consecutive nucleotides are included in the evaluation.

To calculate the thermodynamic interaction of candidate strands (needed for both the TLM and SLM estimators), a modified nearest-neighbor algorithm based on the freely available MELTING4 tool can be used [30]. The MELTING4 code must be modified to handle internal and terminal base pair mismatches [1, 31]. Terminal mismatches are treated by "padding" all evaluated sequences with a complementary 3-bp region. This simulates the motif environment and ensures that the ends are complementary. The padding artificially increases the stability of a configuration (slightly) due to the additional matching base-pairs. This systematic bias means that the calculated values are more reliable as a relative measure of sequence melting temperature than an absolute measure.

Given the vast sequence interaction space that must be covered, the execution time on a single processor can quickly become prohibitively large as the size of the target structure increases. To overcome this limitation the algorithm partitions the problem into subparts, which are then executed in parallel on computing clusters and multiprocessor machines. This greatly reduces the time needed to perform an optimization run and allows the application to target larger and more complex DNA nanostructures. This procedure is called a parallel cross-melt since all physically realistic alignments of strands are evaluated in parallel to determine "cross" reactivity between strands outside of the intended (i.e., designed) interactions.

The following pseudo-code generates a TLM-optimized sticky-end solution arm set for the 20-arm or 96-arm systems:

```
1: FindDNASet(seq_length, set_size, fixed sequence set)
2: {
3:   arms = generate_all_sequences(seq_length);
4:   arms = remove_verboten(arms);
5:   arms = add_complements(arms);
6:   arms = remove_duplicates(arms);
```

```
 7: seq_set = concatenate_set(arms, fixed_sequence_set)
 8: results = cross_melt(seq_set);
 9: results = remove(results, fixed_sequence_set);
10: results = sort_by_TLM(results);
11: top = head(results, set_size);
12: return = top; }
```

The parameters are the target sticky-end length (5 for the cruciform motif), the target sequence set size (20 and 96, respectively) and the fixed sequence set (the A and B cores, shells and arm strands without sticky-ends).

The first step is to generate all possible sequences of the target sticky-end length. In the next step, *verboten* sequences are removed from the problem space if appropriate [22, 25]. Verboten sequences are sequences known to have unfavorable properties for self-assembly in some experimental contexts. The parallel cross-melt algorithm is executed and the resulting interaction data is used to rank the strands based on their specificity (i.e., TLM). The top sequences represent the TLM-optimal solution set.

2.4.2 Alternative Designs

Single Core. The quality of the solution set will improve if the number of fixed motif strands is minimized. This is intuitive since each fixed strand imposes additional constraints on the solution space. For the target system this can be applied by using a single core on all motifs rather than the original dual core motifs (A-type and B-type). The method can optimize for a single core (e.g., A-only or B-only) system by simply modifying the set of fixed sequences dedicated to the motifs (i.e., cores, shells, and arms).

Split Core. Another approach to improve the design method makes use of the two motif types (A and B) in the context of hierarchical assembly. Both the 20-arm and 96-arm systems are assembled in two separate hierarchical steps. In the first step, single strands assemble into motifs. In the second step motifs are mixed and, due to their sticky-ends, assemble into the target grid structure. Thus, the interaction between the sticky-ends and the fixed strands (cores/shells) is only critical in the first level of the hierarchy when motifs are annealed. In the second step (grid-anneal), the motifs are assumed to be thermodynamically stable and the major factor becomes the sticky-end interactions.

This method can generate an optimized set of sticky-ends for each motif type separately and is equivalent to applying the single core method for both A and B motif types. The final set is obtained by combining the results of the two optimizations with any common sequences used only once. The A-type sticky-ends will have suboptimal interaction with the B core/shells (and vice versa), but the hierarchical assembly process guarantees that they are only simultaneously in solution starting with the second assembly step. Thus, the sticky-ends from one type will not interact with the core or shells from the other.

Since the sticky-ends must be unique across the whole system, the effectiveness of this approach depends on the TLM estimates of the resulting solution sets for the A and B motifs. However, the sequences for the A and B core and shell strands were originally designed to be as different as possible in order to minimize their mutual interaction (i.e., a large TLM for each motif type). This is expected to translate into significantly different solution sets for each core.

The pseudo-code for the Split A/B design follows:

```
1: SplitDesign(set_size)
2: {
3:   setA = FindDNASet(5, set_size, A_Core_and_Shells);
4:   setB = FindDNASet(5, set_size, B_Core_and_Shells);
5:   setA = sort_by_TLM(setA);
6:   setB = sort_by_TLM(setB);
7:   for (i = 0;i < (set_size/2);i++)
8:     seq = head(setA, 1);
9:     ret_setA = concatenate_set (ret_setA, seq);
10:    setA = remove(setA, seq);
11:    setB = remove(setB, seq);
12:    seq = head(setB, 1);
13:    ret_setB = concatenate_set (ret_setB, seq);
14:    setA = remove(setA, seq);
15:    setB = remove(setB, seq);
16:  }
17:  return = concatenate_set(ret_setA, ret_setB); }
```

The solution sets for each core are separately computed and the final set is assembled from the top sequences of the two sets.

SEM Optimization. The above methods were presented in the context of obtaining TLM-optimal sequence sets for low assembly error rates. However, SEM could also be an important design goal. For example, the self-assembled system might need to be stable in certain temperature ranges in order to interface with other systems. The method can optionally trade TLM-optimality for SEM-optimality. This trade-off is controlled with an SEM factor (SF) that proportionally expands the candidate set of TLM ranked sequences for subsequent ranking by their SEM estimates. The following pseudo-code illustrates this process:

```
1: FindDNASet(seq_length, set_size, fixed_seq, sem_factor)
2: {
... (lines 3 to 11 from original)
3: ex_set_size = set_size * (1 + sem_factor);
4: top = head (top, ex_set_size);
5: top = sort_by_SEM(top);
6: final = head (top, set_size);
7: return = final; }
```

2.4.3 Design Evaluation

The results of each method (AB, A-only, B-only, and split A/B) can be evaluated in terms of specificity and stability (TLM and SEM estimates) as applied to a small 20-arm system and a structurally similar but larger 96-arm system (as described earlier). The results are compared with the expected values for a random sequence design as well as the original 20-arm set from [18] that was generated by the widely used text-distance tool SEQUIN [25]. Table 2.1 shows the results in terms of SEM, average nonspecific T_m (NST_m), and TLM for each method.

To evaluate the upper bounds for the SEM and TLM of each method all fixed sequences (cores/shells) are removed and the best possible sets for each metric are evaluated. This simulates a theoretical system in which the core and shells do not interact with the sticky-ends. To obtain the highest possible SEM-optimal design a large SF is used, as shown in Table 2.2.

The random sequence method (i.e., 20-arm) results show a fairly high SEM, which translates into good expected stability for the target assembly. However, the overall specificity (TLM) is very low, suggesting that the system is likely to form defective structures when self-assembling.

The SEQUIN-generated 20-arm original design shows a slight increase in both TLM and SEM when compared to random. However, the TLM is still low and this shows that using simple text-distance metrics and heuristics for optimizing sequence sets can lead to uncertain results if low error rates are desired.

The AB core set generated by exhaustive search shows a significant improvement in specificity. The target structure is thus much more likely to form correctly when using this method. The stability estimate is lower than the original design, which means that the system will denature (i.e., diffuse

DNA Self-Assembly 19

Table 2.1
20-Arm and 96-Arm Results

20-Arm	SEM	NST_m	TLM
AB core	7.12	−6.87	9.81
AB core original*	11.77	4.83	4.24
AB random	10.04	4.01	3.32
A-only	7.66	−6.44	9.82
B-only	9.75	−4.68	10.00
B-only SF = 4	14.08	−0.08	9.65
AB split	9.75	−4.74	9.99
AB split SF = 7	15.75	2.32	9.31
96-Arm			
AB core	6.66	−6.65	9.25
AB random	10.04	4.01	3.32
A-only	7.80	−5.83	9.44
B-only	8.17	−5.68	9.52
B-only SF = 1	11.19	−1.38	9.11
AB split	8.28	−5.54	9.57
AB split SF = 7	12.24	−1.29	9.15

* No 96-arm original exists.

Table 2.2
Upper Bounds on TLM and SEM with 5-nt Sticky-Ends

20-Arm	SEM	NST_m	TLM
No core – rank by TLM	11.93	−3.29	10.47
No core – rank by SEM	18.94	6.95	7.22
96-Arm			
No core – rank by TLM	8.96	−5.16	9.77
No core – rank by SEM	15.90	3.39	8.05

apart) at slightly lower temperatures. However, the SF can be used to trade specificity for stability and can be used to balance the design.

Figure 2.3 shows a scatter plot of three sequence sets for the 20-arm system. The diagonal represents the line of zero specificity (TLM = 0) where specific T_m equals nonspecific T_m. The distance to this line from any given point (i.e., sticky-end sequence) is the TLM. The AB core sequences are clustered in a series of points situated at roughly similar TLMs. The random and original designs do not show this pattern and include sequences that are situated on the diagonal itself: these sequences are just as likely to base-pair with the core/shells as they are with their complements! These strands are likely to have a particularly disruptive effect on structure formation in their sets and there is some evidence that this is the case [4].

Figure 2.4 shows that when the number of fixed sequences (cores/shells) decreases (A-only or B-only versus AB cores), the average specificity of the sys-

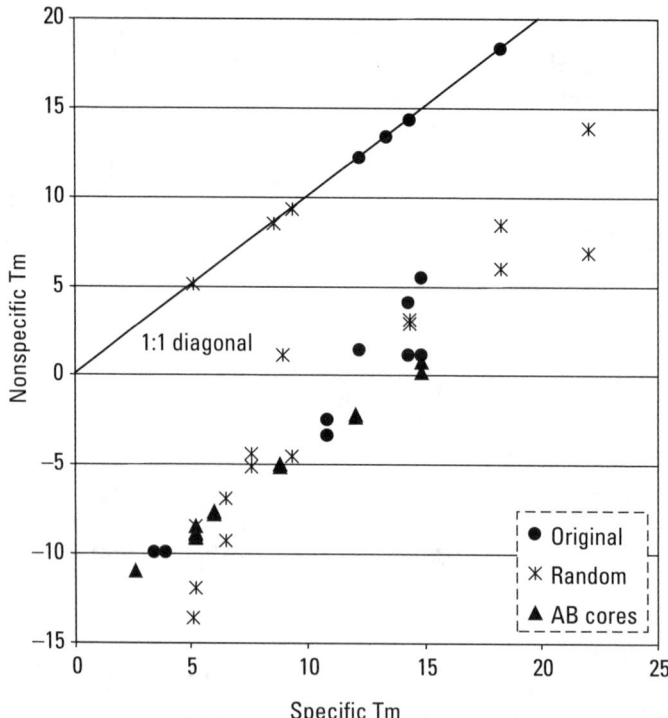

Figure 2.3 Core, random, and original 20-arm sets.

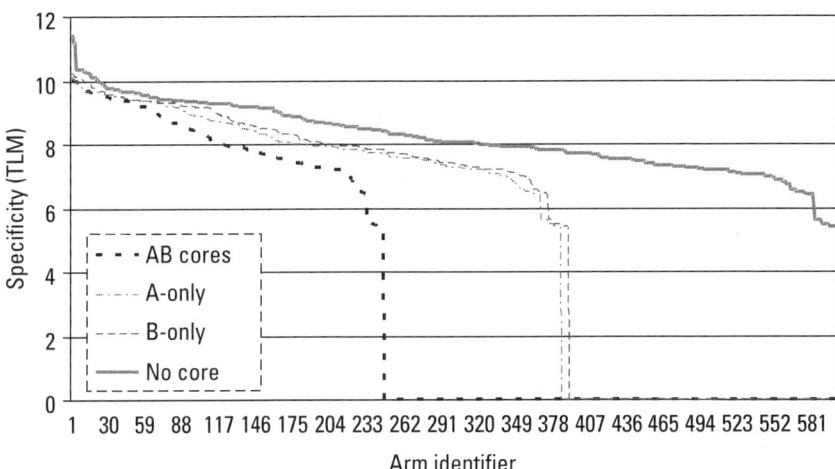

Figure 2.4 Specificity in arm space with different fixed sequence sets.

tem increases. The specificity of each arm is defined as the minimum of its TLM and the TLM of its complement. (In all 5-nt arm systems there are 600 candidate sequences because 524 of the total possible 1,024 sequences are verboten and removed.) This result verifies the intuition that fixed sequences in the motifs restrict the number of high-quality (i.e., high-specificity) arm sequences.

Figure 2.4 also shows the systematic bias induced by the thermodynamic optimization tool due to sequence padding. There is an offset of ~5.4 for nonzero TLM values for all the designs due to the always-complementary 3-bp pads used by the tool.

The results of B-only, AB cores, and the random methods for the 96-arm design are presented in Figure 2.5. The random method has many arms that strongly interact with the core/shells (mapped on the diagonal). B-only shows the same clustering as AB cores but it is on average slightly farther away from the diagonal (i.e., B-only has higher specificity).

The split AB design is the best performer for TLM-optimized designs, slightly outperforming even the single core B-only method for larger systems. Figure 2.6 shows how the SF factor can be used to increase the SEM of a design at the expense of a slightly lower TLM. In Figure 2.7 two such balanced designs, one based on the B-only method and one on an SEM-targeted split AB are contrasted with the text-based design of the original AB and with the random method. The theoretical maximum SEM and TLM

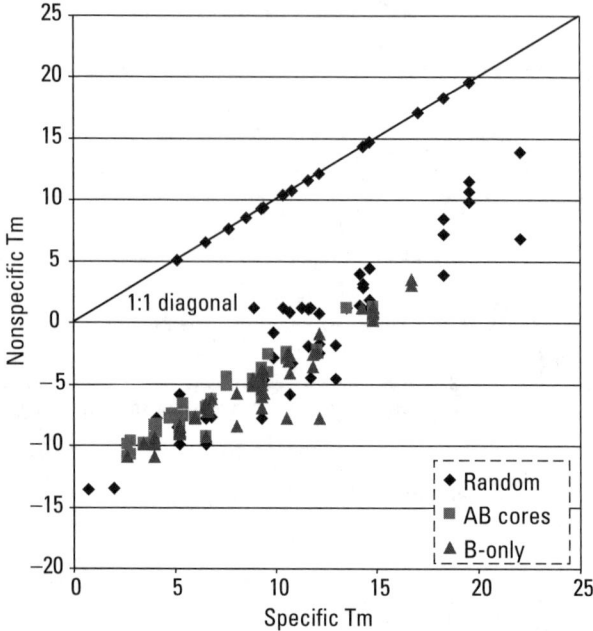

Figure 2.5 B-only, AB cores, and random for 96-arm sets.

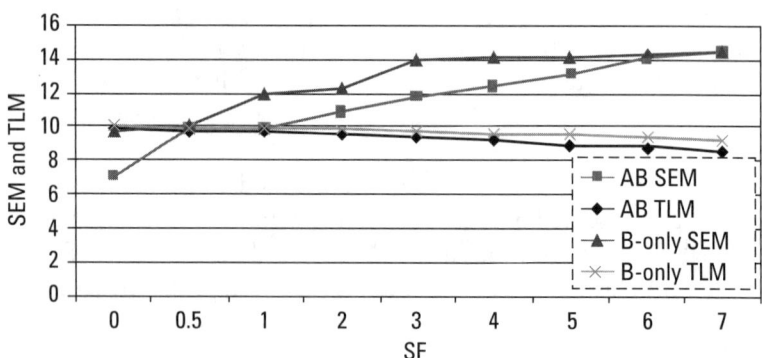

Figure 2.6 Trade-off: specificity can be traded for stability.

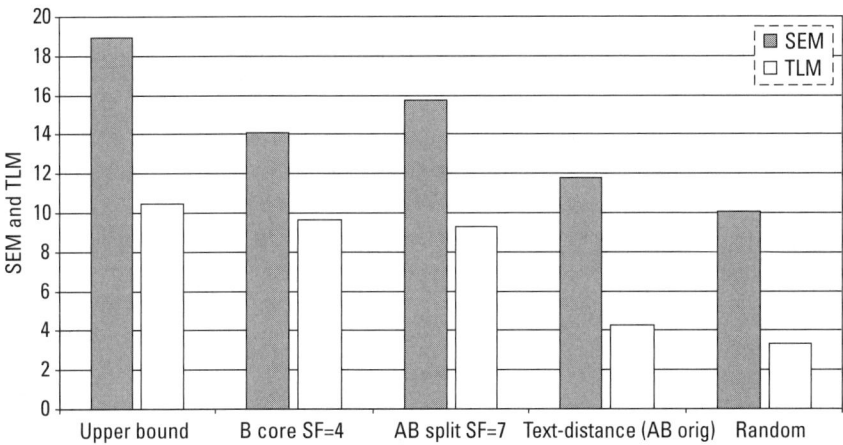

Figure 2.7 20-arm designs balanced for TLM and SEM metrics.

for 5-nt sticky-end designs are included for comparison. Table 2.3 lists the best AB core sticky-end sequences (ordered by SEM).

2.5 Hierarchical Assembly

The method described above can be extended to create aperiodic DNA self-assembled nanostructures [4, 5, 32] with four uniquely edged 4 × 4 grid structures. Since each motif is assembled from five common and four unique oligonucleotides in an individual vessel, each motif can be independently modified and can create arbitrary patterns, as shown in Figure 2.8. The grids are selectively functionalized with the protein streptavidin by using a biotin-functionalized core strand during the annealing of the motifs and by introducing free streptavidin after assembly.

The cost of this technique scales with the number of unique oligos in the final nanostructure (e.g., 69 oligos in this case). The origami method [10, 33] suffers from this scaling law as well but in principle could use the methods outlined here to reduce the cost of larger multi-shape structures.

Regardless of the method employed to assemble the basic motif, a unique set of sticky-ends is required to assemble a structure unambiguously and the reuse of strands is needed to decouple the cost of a structure from the linear dependence on its area. Thus, such a method is scalable in terms of the size of the nanostructure that can be fabricated from a finite DNA sequence

Table 2.3
The Best 160 × 5-nt Arm Sequences (and Complements) Found by the AB-Core Method

Arm	Comp.	Arm	Comp.	Arm	Comp.
CGTGC	GCACG	CCTCG	CGAGG	TATGT	ACATA
CAAGC	GCTTG	ACGAC	GTCGT	TGTAT	ATACA
ACGTC	GACGT	CAGAC	GTCTG	TTAGA	TCTAA
ACAGC	GCTGT	ACTGC	GCAGT	TTACT	AGTAA
TGCAG	CTGCA	TGCTG	CAGCA	TAAGA	TCTTA
CTGTG	CACAG	TGCAC	GTGCA	AATAG	CTATT
AGCTC	GAGCT	AGAGC	GCTCT	AATTC	GAATT
CATGG	CCATG	CTTGG	CCAAG	ATACT	AGTAT
CAATC	GATTG	CATTC	GAATG	TAACT	AGTTA
AATGC	GCATT	ATTGC	GCAAT	TTAGT	ACTAA
AACGT	ACGTT	CTAAC	GTTAG	TACTT	AAGTA
CTTAC	GTAAG	TTACG	CGTAA	TAAGT	ACTTA
TAACG	CGTTA	CATTG	CAATG	TAGAT	ATCTA
ATGAC	GTCAT	ATGCT	AGCAT	TAGAC	GTCTA
TCATG	CATGA	TTGCT	AGCAA	TTCAT	ATGAA
TTGAG	CTCAA	AAGCT	AGCTT	ATTCT	AGAAT
TGCTT	AAGCA	ACTGT	ACAGT	TCAAT	ATTGA
TCACA	TGTGA	AGTAC	GTACT	TTAAC	GTTAA
TACGT	ACGTA	TGTAG	CTACA	AATCT	AGATT
TACTG	CAGTA	AAGTG	CACTT	TGATT	AATCA
TCAAC	GTTGA	TCTGA	TCAGA	TCATT	AATGA
TCTAG	CTAGA	TAGCT	AGCTA	TATGA	TCATA
AATGT	ACATT	ATGTT	AACAT	TTAAG	CTTAA
ATTGT	ACAAT	TTCAA	TTGAA	TTACA	TGTAA
GTTAT	ATAAC	AATAC	GTATT	TTGTA	TACAA
TAATG	CATTA	CAATA	TATTG	TTGAT	ATCAA
TTATG	CATAA	AATTG	CAATT		

These sticky-end sequences are compatible with the tile motif and system described in [4–6].

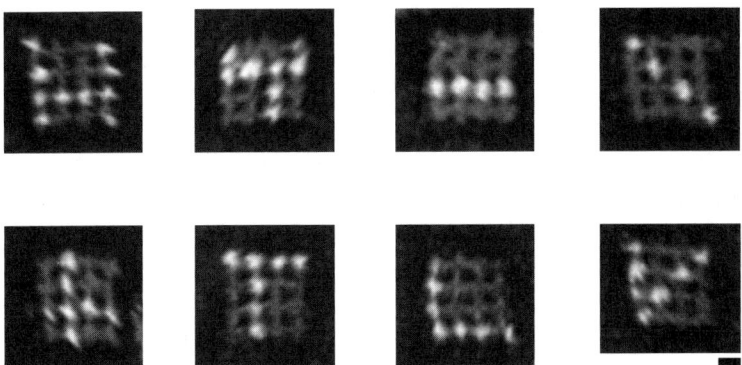

Figure 2.8 AFM images of streptavidin patterned 4 × 4 DNA grids. Scale bar is 20 nm.

space. In the limit, a single oligonucleotide sequence might be used to form large supramolecular structures [34]. However, in order to retain maximal programmability, multi-strand designs with a variable degree of reuse should be used. Strand reuse is exploited to some extent in both methods [5, 10, 33], but this had not previously been demonstrated as a viable alternative for assembling large aperiodic structures prior to the introduction of the hierarchical method.

There are at least two related methods to achieve scalable DNA self-assembly. The two methods described here build larger structures from small motifs in a hierarchical manner. Since there are a finite number of good sticky-end sequences for a given design (e.g., 5-nt sticky-ends), each method employs strand reuse to overcome this natural limitation on structure size.

2.5.1 Generic Linkers

The first approach uses a series of "generic" sticky-ends along the periphery of a DNA grid to aid in binding a grid to an adjacent grid. The generic linkers[1] are designed to bind with only one helix (typical arms have two helices) to introduce a relatively unstable interaction. This is to prevent the generic linkers from dominating the specific interactions that will be introduced later to programmably organize two distinct grids. The generic linkers are replicated along the grid edge to liberate the few otherwise specific sequences and

---

[1] The left linker in is 5'-TAGATGATAGAGTGGTACATCT-3' and the right is 5'-ATCTAACGGATGAGTAGTGGGCTCAGTCGGAT -3'.

thereby enable stable binding between two grids. It is then possible to apply an incremental graph-coloring method, with a constant number of specific binding sites, to sequentially add motifs to the growing structure [35].

The specific binding of the two distinct grids is achieved by using nongeneric (i.e., specific) sticky-ends at selective locations—in this example at the lower right and upper right corners of the two 4 × 4 grids, respectively (this will become the middle-right edge in the 32-motif structure). Figure 2.9 shows the relationship between the generic and specific arms. The gap along the edge between the two 4 × 4 grids is introduced to disambiguate the identity of each grid in the assembly after deposition on a surface.

This demonstrates that multiple weak interactions between oligos along the edge of a grid (generic linkers) can be controlled (or dominated) by a single strong specific interaction. Moreover, this is evidence that the free energies among distal nucleotide interactions constructively add between bound motifs. The gap seen in the structure from Figure 2.9 was the previously described mark designed for identification purposes and may disrupt cumulative nonspecific binding induced by the generic oligos. The stability (at room temperature) of such assemblies is compromised because of the weakness of the generic linkages and gap. This may contribute to a reduced apparent yield of the 32-motif structure. However, this demonstration illustrates that the use of generic linkers along the edge of a DNA nanostructure can be dominated by a single specific interaction and enables the development of scalable sequential assembly methods.

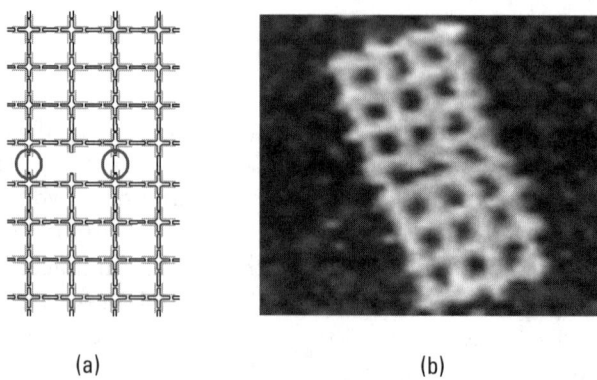

(a) (b)

Figure 2.9 (a) Schematic and (b) AFM image of a two-grid (4 × 8) assembly. Only one linkage at the interface between the two distinct 4 × 4 grids is specific (right-most) and the remaining two are generic (circled). Scale bar is 40 nm.

2.5.2 Fully Specific Linkers

The second method uses fully specific sticky-ends along the periphery of each subgrid. The reuse of sticky-end sequences from within each grid reduces the number of unique oligos required to assemble larger arrays. This does not induce ambiguous assembly because the sticky-ends are reused after the constituent pieces of the grids have already formed. For example, sticky-ends from the four tetramers in each grid can be reused after the grid has assembled. This is only possible because strand exchange between the reused sticky-ends and the intra-grid sticky-ends does not occur frequently if at all.

Sticky-end reuse can be tested by assembling a 2 × 2 array of grids. Figure 2.10 illustrates a typical AFM scan of the assembled 64-motif (8,960 kD) product on cleaved mica. The method used to pattern the 4 × 4 grids from Figure 2.8 is also applicable here.

The use of an AFM image to determine the "yield" of assembly is a not an accurate method since the mica surface used as a substrate (with the imaging buffer) will preferentially bind large, flat, charged DNA structures. Further, the simple motifs used here, unlike those in [10], are not easily imaged alone by AFM but must be assembled into larger structures to be reliably observed.

Thus, to determine the relative merit of the technique, the defect probability of assembled structures must be evaluated since the "raw" motif yield

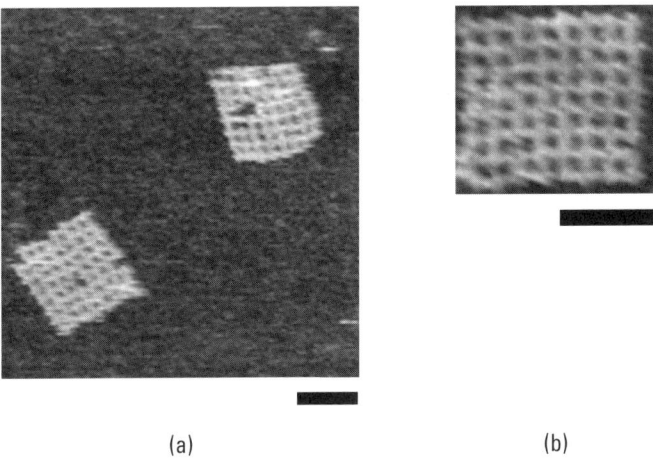

(a)   (b)

Figure 2.10 (a) An AFM scan of the 2 × 2 array of grids; (b) the assembled array demonstrates stability even under repeated AFM scans. Scale bars are 100 nm.

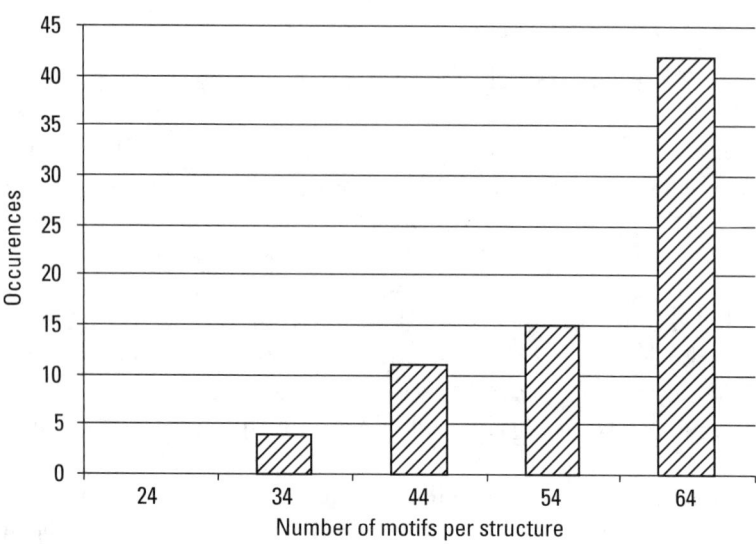

Figure 2.11 Histogram of identifiable structures as observed by AFM ($N = 73$).

is unknown. Figures 2.11 and 2.12 are histograms of the AFM-observed, surface-bound assembly products. Structures are classified as either *identifiable* (i.e., as a 64-motif grid, $N = 73$) or *unidentifiable* (i.e., fragments, $N = 43$).[2]

The binning used in Figures 2.11 and 2.12 is 10 and 5 motifs wide, respectively. For example, incomplete structures with 55 motifs will be binned with fully intact structures with 64 motifs in Figure 2.11. Structures were disregarded if any of the following was observed: (1) clipping by the scan window; (2) piling into agglomerations on the surface; or (3) manipulation away from the surface during a line scan.

The presence of non-ideal, defective structures is not surprising. As noted in [6], the availability of purification methods is a key advantage of DNA self-assembly in the face of defects. The use of solid-support or affinity binding purification may be able to remove defective structures. The *yield-scalability* of this technique will depend on the efficiency with which defective material can be removed. However, the *size-scalability* of the technique depends only on sequence and motif reuse. Recent work in error-resilient nanostructure design may also be applicable within this framework [36–40].

---

[2]N is the number of structures in each category.

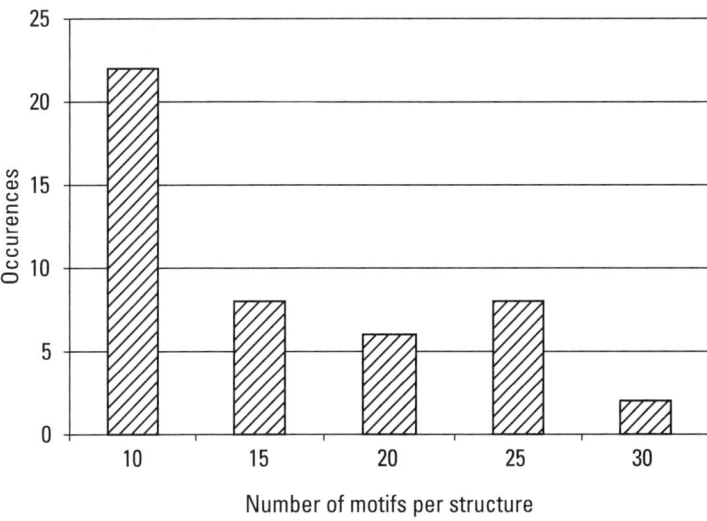

Figure 2.12 Histogram of unidentifiable structures as observed by AFM (*N* = 43).

2.6 Summary

The continued scaling of conventional CMOS fabrication techniques faces many challenges that may be overcome by a switch to bottom-up self-assembly. DNA's precise binding rules at small scale (a few nanometers) make it a potential candidate for future fabrication of nanoscale computing systems. This chapter presented a thermodynamics-based, hierarchical design methodology for creating cost-effective, scalable DNA nanostructures with aperiodic 2D patterns. This capability forms the foundation upon which we may fabricate DNA self-assembled computer systems.

References

[1] Peyret, N., et al., "Nearest-Neighbor Thermodynamics and NMR of DNA Sequences with Internal A-A, C-C, G-G, and T-T Mismatches," *Biochemistry*, Vol. 38, No. 12, 1999, pp. 3468–3477.

[2] Whitesides, G. M., and B. A. Grzybowski, "Self-Assembly at All Scales," *Science*, Vol. 295, 2002, pp. 2418–2421.

[3] Wang, X., and W. M. Nau, "Kinetics of End-to-End Collision in Short Single-Stranded Nucleic Acids," *Journal of the American Chemical Society*, Vol. 126, 2004, pp. 808–813.

[4] Dwyer, C., et al., "The Design and Fabrication of a Fully Addressable 8-tile DNA Lattice," *Proc. Foundations of Nanoscience: Self-Assembled Architectures and Devices*, 2005, pp. 187–191.

[5] Park, S. H., et al., "Finite-Size, Fully-Addressable DNA Tile Lattices Formed by Hierarchical Assembly Procedures," *Angewandte Chemie*, Vol. 45, 2006, pp. 735–739.

[6] Pistol, C., and C. Dwyer, "Scalable, Low-Cost, Hierarchical Assembly of Programmable DNA Nanostructures," *Nanotechnology*, Vol. 18, 2007, pp. 125305-9.

[7] Chworos, A., et al., "Building Programmable Jigsaw Puzzles with RNA," *Science*, Vol. 306, No. 5704, 2004.

[8] Koyfman, A. Y., et al., "Controlled Spacing of Cationic Gold Nanoparticles by Nanocrown RNA," *Journal of the American Chemical Society*, Vol. 127, No. 34, 2005, pp. 11886–11887.

[9] Pistol, C., A. R. Lebeck, and C. Dwyer, "Design Automation for DNA Self-Assembled Nanostructures," *Proc. 43rd Design Automation Conference (DAC)*, 2006, pp. 919–924.

[10] Rothemund, P. W. K., "Folding DNA to Create Nanoscale Shapes and Patterns," *Nature*, Vol. 440, No. 7082, 2006, pp. 297–302.

[11] Seeman, N. C., "Nucleic Acid Junctions and Lattices," *Journal of Theoretical Biology*, Vol. 99, 1982, pp. 237–247.

[12] Seeman, N. C., "DNA Engineering and Its Application to Nanotechnology," *Trends in Biotechnology*, Vol. 17, 1999, pp. 437–443.

[13] Goodman, R. P., et al., "Rapid Chiral Assembly of Rigid DNA Building Blocks for Molecular Anofabrication," *Science*, Vol. 310, No. 5754, 2005, pp. 1661–1665.

[14] Seeman, N. C., et al., "The Perils of Polynucleotides: The Experimental Gap Between the Design and Assembly of Unusual DNA Structures," *Proc. Second International Meeting on DNA Based Computers (DNA2)*, 1996, pp. 191–205.

[15] Mao, C., W. Sun, and N. C. Seeman, "Designed Two-Dimensional DNA Holliday Junction Arrays Visualized by Atomic Force Microscopy," *Journal of the American Chemical Society*, Vol. 121, 1999, pp. 5437–5443.

[16] Sharma, J., et al., "DNA-Templated Self-Assembly of Two-Dimensional and Periodical Gold," *Angewandte Chemie–International Edition*, Vol. 45, No. 5, 2006, pp. 730–735.

[17] Reishus, D., et al., "Self-Assembly of DNA Double-Double Crossover Complexes into High-density, Doubly Connected, Planar Structures," *Journal of the American Chemical Society*, Vol. 127, No. 50, 2005, pp. 17590–17591.

[18] Yan, H., et al., "DNA Templated Self-Assembly of Protein Arrays and Highly Conductive Nanowires," *Science*, Vol. 301, 2003, pp. 1882–1884.

[19] He, Y., et al., "Self-Assembly of Hexagonal DNA Two-Dimensional (2D) Arrays," *Journal of the American Chemical Society*, Vol. 127, No. 35, 2005, pp. 12202–12203.

[20] Le, J. D., et al., "DNA-Templated Self-Assembly of Metallic Nanocomponent Arrays on a Surface," *Nano Letters*, Vol. 4, No. 12, 2004, pp. 2343–2347.

[21] Zheng, J. W., et al., "Two-Dimensional Nanoparticle Arrays Show the Organizational Power of Robust DNA Motifs," *Nano Letters*, Vol. 6, No. 7, 2006, pp. 1502–1504.

[22] Winfree, E., et al., "Design and Self-Assembly of Two-Dimensional DNA Crystals," *Nature*, Vol. 394, 1998, pp. 539–544.

[23] Barish, R. D., P. W. K. Rothemund, and E. Winfree, "Two Computational Primitives for Algorithmic Self-Assembly: Copying and Counting," *Nano Letters*, Vol. 5, No. 12, 2005, pp. 2586–2592.

[24] Liu, D., et al., "DNA Nanotubes Self-Assembled from TX Tiles as Templates for Conductive Nanowires," *Proceedings of the National Academy of Science*, Vol. 101, 2004, pp. 717–722.

[25] Seeman, N. C., "De Novo Design of Sequences for Nucleic Acid Structural Engineering," *Biomolecular Structure & Dynamics*, Vol. 8, No. 3, 1990, pp. 573–581.

[26] Hartemink, A. J., D. K. Gifford, and J. Khodor, "Automated Constraint-Based Nucleotide Sequence Selection for DNA Computation," 1998, pp. 227–235.

[27] Feldkamp, U., et al., "DNASequenceGenerator: A Program for the Construction of DNA Sequences," *Proc. Seventh International Workshop on DNA Based Computers (DNA7)*, 2001, pp. 23–32.

[28] Yin, P., et al., "TileSoft: Sequence Optimization Software for Designing DNA Secondary Structures," Technical Report CS-2004-09, Dept. of Computer Science, Duke University, 2004.

[29] Shortreed, M. R., et al., "A Thermodynamic Approach to Designing Structure-Free Combinatorial DNA Word Sets," *Nucleic Acids Research*, Vol. 22, No. 15, 2005, pp. 4965–4977.

[30] Le Novere, N., "MELTING, Computing the Melting Temperature of Nucleic Acid Duplex," *Bioinformatics*, Vol. 17, 2001, pp. 1226–1227.

[31] Santa Lucia, J., Jr., and D. Hicks, "The Thermodynamics of DNA Structural Motifs," *Annual Review of Biophysics and Biomolecular Structure*, Vol. 33, 2004, pp. 415–440.

[32] Lund, K., et al., "Self-Assembling a Molecular Pegboard," *Journal of the American Chemical Society*, Vol. 127, No. 50, 2005, pp. 17606–17607.

[33] Rothemund, P. W. K., "Design of DNA Origami," *Proc. IEEE/ACM International Conference on Computer Aided Design (ICCAD)*, 2005, pp. 471–478.

[34] Liu, H. P., et al., "Approaching the Limit: Can One DNA Oligonucleotide Assemble into Large Nanostructures?" *Angewandte Chemie–International Edition*, Vol. 45, No. 12, 2006, pp. 1942–1945.

[35] Dwyer, C., et al., "The Design of DNA Self-Assembled Computing Circuitry," *IEEE Transactions on VLSI*, Vol. 12, 2004, pp. 1214–1220.

[36] Winfree, E., and R. Bekbolatov, "Proofreading Tile Sets: Error Correction for Algorithmic Self-assembly," *DNA Computing Lecture Notes in Computer Science*, Vol. 2943, 2004, pp. 126–144.

[37] Chen, H. L., and A. Goel, "Error Free Self-Assembly Using Error Prone Tiles," *DNA Computing Lecture Notes in Computer Science*, Vol. 3384, 2005, pp. 62–75.

[38] Soloveichik, D., and E. Winfree, "Complexity of Compact Proofreading for Self-Assembled Patterns," *DNA Computing Lecture Notes in Computer Science*, Vol. 3892, 2006, pp. 305–324.

[39] Reif, J. H., S. Sahu, and P. Yin, "Compact Error-resilient Computational DNA Tiling Assemblies," *DNA Computing Lecture Notes in Computer Science*, Vol. 3384, 2005, pp. 293–307.

[40] Baryshnikov, Y., et al., "Self-Correcting Self-Assembly: Growth Models and the Hammersley Process," *DNA Computing Lecture Notes in Computer Science*, Vol. 3892, 2006, pp. 1–11.

# 3

Self-Assembled Circuitry and Design

3.1 Introduction

This chapter describes techniques for using DNA self-assembly for the implementation of molecular-scale circuits. First, a brief survey of the most-closely related work is provided. This is followed by a discussion of the two primary methods for creating DNA self-assembled circuits: (1) DNA scaffolding structures, and (2) DNA-guided self-assembly. The chapter concludes with a description of a design tool flow and some case studies of circuit designs.

3.2 Related Work

3.2.1 Mesoscale Self-Assembly

Mesoscale is a term used to describe the size regime from microns to millimeters. Self-assembly at mesoscopic lengths must employ specific interactions that are stronger than the interfacial interactions between particles. The interfacial energy can be large and thus overpower delicate interactions such as single stranded DNA hybridization. However, bundles (or brushes) of DNA can be used to assemble particles at this scale but with much less specificity due to the many stable non-Watson-Crick base pairing configurations (e.g., nonspecific base stacking, surface adsorption, and so on) that are available. Surface energy and hydrophilic or hydrophobic interactions are more suitable for mesoscale assembly since they are strong and can overcome nonspecific interactions.

However, mesoscale assembly mechanisms lack the richness of single-molecule interactions used by proteins or DNA. Thus, mesoscale assemblies must leverage some other form of information content to reliably and controllably create interesting structures. One such method, called microelectromechanics (MEMs), employs top-down patterning to place microscale beams and paddles (with or without circuitry) sufficiently close to each other and in the correct pattern to enable the folding of the pieces into a final structure after being released from the surface. The folding mechanism can be driven by any number of phenomena including surface free energy minimization, thermal expansion, or even high-frequency vibrations from outside of the system.

3.2.2 Nanoparticle-DNA Conjugates

At the intersection between the molecular scale and the mesoscale, DNA conjugates retain much of their single-molecule selectivity without being encumbered by weakness toward nonspecific interactions. In such cases DNA can be used to precisely bind nanoparticles, such as spheres, rods, or triangular solids in interesting configurations. One of the first systems to demonstrate this approach used gold colloids and complementary, thiol-functionalized DNA strands to create clusters [1]. While the DNA successfully bound two distinct gold particles together, the approach does not scale to complex, asymmetric structures due to the inherent symmetry of the gold particle. Efforts to break the symmetry of the gold particle by curvature-sensitive self-assembled monolayers on the particle surface are beginning to emerge as viable new approaches to create asymmetric colloidal structures [2].

3.3 DNA Scaffolding Structures

DNA can be used to assemble large 2D grids by using the hierarchical methods outlined previously. Grids composed of hundreds of thousands of individual motifs and extending up to at least 10 microns on their long edge have been created [3, 4]. These DNA nanostructures can provide a scaffold onto which molecular-scale devices can be attached.

As described above, nanoparticles can be functionalized with single strands of DNA. This nanoparticle can be placed at a specific location if the DNA scaffold is designed to expose the complementary strand at a specific location. The resolution of this location is partially determined by the DNA scaffold design and partly by the diameter and number of strands on the nanoparticle. The DNA scaffold described in Chapter 2, with repeating cavities of ~16 × 16 nm and a 4-nm separation between cavities, will be used here [5–7]. The goal of this chapter is to explore how this scaffold can be used to create nanoscale circuits.

DNA scaffolding is a general technique independent of the specific nanoelectronic device to be placed on the scaffold. Many of the techniques presented in this chapter are also device agnostic. That is, the techniques described here can apply to any number of other suitable device structures. However, to make the circuit discussions concrete, the focus here is on carbon nanotubes devices.

3.4 Implications for Nanoelectronic Circuit Architecture

To use DNA scaffolds, the nanoelectronic circuit architecture must strike a balance between (1) the regularity of DNA self-assembly patterning capabilities, (2) the complexity required for sophisticated system designs, and (3) tolerance to the inevitable defects present in nanoscale systems. The remainder of this section elaborates on each of these issues and focuses on the fundamental differences between this nanoscale circuit architecture and current CMOS-based circuits.

3.4.1 Regularity

While the design of CMOS-based circuits can be simplified by the use of regularity (e.g., standard cell design), regularity is not a fundamental requirement. Through the use of unique tags DNA can form regular or arbitrary structures. However, DNA self-assembly has a potential limitation in that the probability of incorrect tag matches increases as the number of unique tags increases. Each type of connection requires a unique pair of complementary single-stranded DNA (ssDNA) tags. With more types of connections and a fixed number of base-pairs per tag, the tags become more similar (i.e., differ in fewer base-pairs) and partial matches become more likely. For example, if a functionalized nanotube binds to a partially matched tag then it is in the wrong position. This situation is analogous to the Hamming distance between encodings of symbols; encoding more symbols with the same number of bits makes the Hamming distance smaller and the probability of an error greater. Minimizing the number of tags reduces the chances of partial matches, which could cause positional defects during annealing. Therefore, repetitive structures are desirable and circuit and system designers should strive to use them as much as possible.

3.4.2 Complexity

Design complexity is a function of the number of different component types and the placement of these components. Current CMOS-based circuits can place hundreds of millions of devices and wires with precision on the order of 45 nm. This precision is achieved by using photolithography to specify exactly where each individual component belongs. With the combination of carbon nanotube devices and DNA self-assembly the goal is to develop circuits that are complex enough to perform interesting computation. The number of component types can be limited to just carbon nanotube field-effect transistors (CNFETs), nanotube wires, and metal plating for connecting wires. However, with DNA self-assembly, component placement cannot be specified with the same degree of complexity as with CMOS. For example, the cost of a CMOS design does not scale with *how* components are placed but rather with the area they occupy.

Complexity in DNA self-assembly must be introduced without requiring a large number of tags. This mirrors the desire to use regular structures that minimize the degree of uniqueness and complexity in a design. However, regular structures typically limit the functionality of circuits.

Thus, the utility of self-assembled DNA arrays depends on the amount of complexity that can be introduced at various levels without causing an intractable number of partial matches. Consider the graph generated from the netlist for a transistor-level design of a combinational circuit. The vertices are transistor terminals and the edges are wires connecting the device terminals. A two-input CMOS NAND gate has about 10 vertices and a useful circuit might require several thousand NAND gates, or 10s of thousands of vertices. Clearly, the naïve approach of assuming a unique tag for each vertex in the graph requires a large number of unique tags (even ignoring fan-out). This will cause too many partial matches that can create faults that render the circuit useless.

3.4.3 Defect Tolerance

A defect is a permanent physical fault that was introduced during fabrication. Consider two types of defects: functional and positional. A functional defect corresponds to a component that does not perform its specified function (e.g., a transistor that does not conduct when it should). A positional defect corresponds to a (functionally correct) component that is placed incorrectly. Both CMOS and DNA self-assembled nanoelectronics can incur functional defects, but only self-assembly is likely to incur positional defects. Positional defects can be defects of both omission and commission. An omissive positional defect occurs when a component is not placed where it belongs. A commissive positional defect occurs when a component is placed where it does not belong (i.e., the partial match described above). Omissive defects behave similar to functional defects. Commissive defects are more dangerous, since they can behave like bridging faults. For example, a misplaced nanotube could cause a short between power and ground or it could change circuit functionality in unpredictable ways (e.g., by erroneously connecting the output of a gate to its input).

In CMOS-based circuits, there is limited support for defect tolerance because of little necessity. The photolithographic placement of components is a mature technology that incurs few defects. However, in architectures with hundreds of millions to billions of devices and wires, defects will occur (i.e., yield is less than 100%). In fact, even now CMOS microchips are tested for defects. If a defect is uncovered and it cannot be tolerated, the chip is discarded. However, some limited number of defects can be tolerated. For example, a defect in a cache or memory cell can be tolerated by systems that provide redundant cells and allow for remapping, or sparing.

Functional and positional defect densities for carbon nanotube devices and DNA assembled nanoelectronics are currently unknown. However, functional defect tolerance can be achieved with the same techniques used in CMOS, since the problem is fundamentally the same. However, tolerance of commissive positional defects is a new challenge. Since positional defect densities are unknown the approach here first strives to minimize positional defects by exploiting regularity in DNA self-assembly. However, as complexity increases and regularity is diminished, the probability of positional defects increases. Thus, more sophisticated circuitry will require more defect tolerance.

3.5 Nanoelectronic Circuit Building Blocks

To establish a point of reference, let us assume a circuit architecture based on a grid of crossed carbon nanotube field effect transistors (CNFETs) with metallic carbon nanotubes (CNTs) for interconnect as shown in Figure 3.1. This process assumes the availability of highly enriched metallic and semiconductor carbon nanotube source materials and a method to functionalize the nanotubes with DNA. At a high level, the design from Figure 3.1 addresses the conflicting goals of regularity and complexity by placing identical unit cells in the cavities of an aperiodic patterned DNA grid. The grid is regular in structure but has aperiodic binding points at which unit cells can be

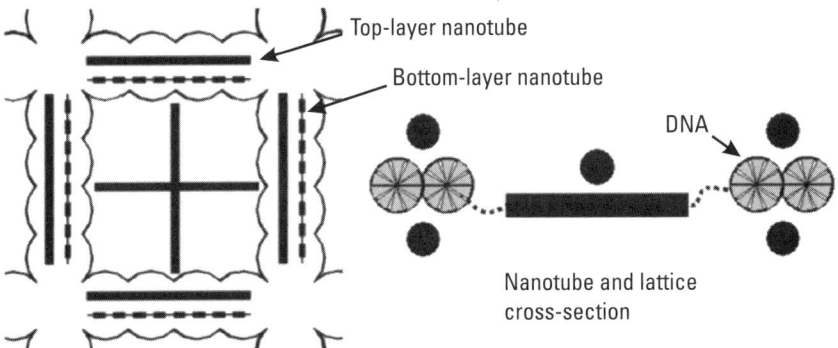

Figure 3.1 Schematic and cross-section of the DNA grid scaffold.

connected into complex patterns. This highlights a key difference between these DNA self-assembled circuits and alternative approaches. Current nanoelectronic architectural approaches assume regularity in both the structure and interconnect network.

3.5.1 Exploiting Regularity: A Replicated Unit Cell

The unit cell in this design is a three-terminal CNFET sitting in the cavity of a DNA grid. A CNFET is placed into the cavity by binding its ssDNA tags to complementary tags on the cavity edges. Semiconducting nanotubes and one metallic nanotube are functionalized such that they bind to the correct tags to form a cross. It is assumed that one of the nanotubes is wrapped in a thin dielectric layer, such as SiO_2, to form a gate oxide [8]. The metallic nanotube functions as the gate of the CNFET and the semiconducting nanotube as the channel.

Using carbon nanotubes of a short length (~16 nm) will preclude commissive positional defects in which a carbon nanotube might bind in two different cavities. By using two sets of tags in alternating cavities along each dimension, a nanotube of a precise length cannot span cavities to form a defect. The distance between adjacent cavities is only 4 nm, so if the same tag is used in adjacent cavities, a nanotube may bind across the arm rather than within a cavity. Using a checkerboard pattern of alternating tags, with sufficient Hamming distance, eliminates positional commissive defects. However, this approach requires carbon nanotubes of a precise length, which may be possible using a sonochemical method and size-exclusion chromatography to separate the nanotubes by their length [9]. This technique must be applied to both the semiconducting and metallic nanotubes after enrichment.

The unit cell is augmented with short metallic carbon nanotubes that lie adjacent to the cavity on both the top and bottom of the DNA grid. The short nanotubes are far enough apart to avoid cross-talk or bridging faults and may also be wrapped with a dielectric if necessary. The nanotubes initially do not intersect to form complete circuits. Instead, an electrical connection between nanotubes must be explicitly created by specifying an appropriate tag on the DNA grid to which a metallic nanosphere will bind. The nanosphere nucleates metal ions to form the connection with the help of an electroless plating process [10, 11]. Similarly, connecting transistors may require specifying whether the device connects to the top or bottom metallic nanotube. Forming these connections is where complexity is added to the design through the use of the methods described in the previous chapter.

This unit-cell design fosters regular, repetitive structures. All nanotubes are the same length (16 nm) and there are five sets of nanotubes that are functionalized with different tags; Four sets of nanotubes are used for the CNFETs: two semiconducting sets and two metallic sets. This corresponds to the two tag sets of the checkerboard pattern of cavity tags. A nanotube from one set can bind to any cavity with the complementary tag. Similarly, the interconnect nanotubes (the fifth set) can bind adjacent to any cavity directly on either the top or bottom of the DNA grid in either the vertical or horizontal direction. This approach enables the use of a regular pattern for the underlying DNA scaffold.

3.5.2 Introducing Complexity: An Aperiodic Pattern for Interconnecting Cells

The building block, while regular in structure, has aperiodic binding points for connecting together the nanotubes of the unit cell. Aperiodic patterning can be achieved using the hierarchical methods outlined in the previous chapter.

Complex circuits can now be constructed by specifying the electroless plating points on the DNA grid to electrically fuse placed nanotubes. For the top and bottom of the grid the plating point options include: (1) the three transistor terminals to a CNFET, (2) those to interconnect nanotubes in the vertical directions, and (3) those to interconnect nanotubes in the horizontal directions. Assume that creating a straight-through connection in the vertical direction requires both of the vertical (i.e., the "north" and "south") connections; similarly, both the east and west connections are required for a straight connection in the horizontal direction. Pass-throughs from the top-level interconnect to the bottom-level are created by connecting a transistor terminal to both interconnects.

Only a single tag on the DNA grid is required to specify the plating points where the metallic nanospheres can bind on the grid. It is this tag that has the aperiodic pattern and nanospheres will bind only where the tag appears. This approach minimizes positional defects since the nanotubes are precise in length and can only bind in the appropriate positions on the grid. In contrast, if a long or random-length nanotube is used to connect distant points it might incorrectly bind to tags along an arbitrary circumference. Thus, length control prevents positional defects.

3.6 Large-Scale Interconnection of Circuit Nodes

The computational capabilities of the proposed building block, called a node, are limited by the size of the DNA grid. Increasing the computing capacity requires interconnecting multiple building blocks. By using inexpensive laboratory equipment it is possible to simultaneously self-assemble as many as 10^{12} identical, but small, nodes. This number of nodes, if placed on a 2-micron pitch, would cover a 175 × 175-cm area, or the equivalent area of ~40 × 300-mm wafers.

Although the size of an individual node is well above the minimum feature size of photolithography (<45 nm), the large number of nodes fabricated through self-assembly restricts conventional techniques. Self-assembling nodes onto a substrate at well-defined places is also difficult without "naming" each placement site (pick and place methods will not scale to this number of components). Even with DNA tags or chemical patterns on the substrate, the nodes are not guaranteed to fall into place precisely.

Most conventional architectures require precise placement and interconnection between circuits. Therefore, even with conventional photolithographic-interconnect node networks, the result would be a random interconnection due to the random placement of nodes on the substrate. This is the sacrifice a self-assembly process imposes: precision and control exist only at small length scales (e.g., less than 10 microns, for now).

One solution to this problem involves a large-scale self-assembling process that can interconnect nodes already placed on a substrate using another form of DNA self-assembly. Individual DNA strands self-assemble between node edges, providing a scaffold for metal that forms an electrical connection, which has been experimentally demonstrated [12, 13]. This larger scale process cannot deliver the precise control found in the earlier process used to assemble the nodes, but it can fabricate single wire interconnections between the edges of the nodes, as illustrated in Figure 3.2.

3.7 DNA Design Flow

Designing grid circuitry requires various custom design tools to facilitate the new technology—for example, tools to specify where nanotubes assemble onto the grid and their connectivity. Since placement and routing are provided by DNA self-assembly, these processes require design tools that can choose DNA tags (i.e., which sequence of base pairs) for the scaffolding and for functionalizing the nanotubes. Unlike existing design tools for silicon CMOS technology, the tools must produce DNA tag sequences instead of mask artwork.

This section describes a tool flow developed for the design of self-assembled circuitry. Sections 3.7.2 to 3.7.4 describe the methods and illustrate where custom tools must be introduced to handle the new technology. Section 3.8 describes the application of the design tools to several logic circuits.

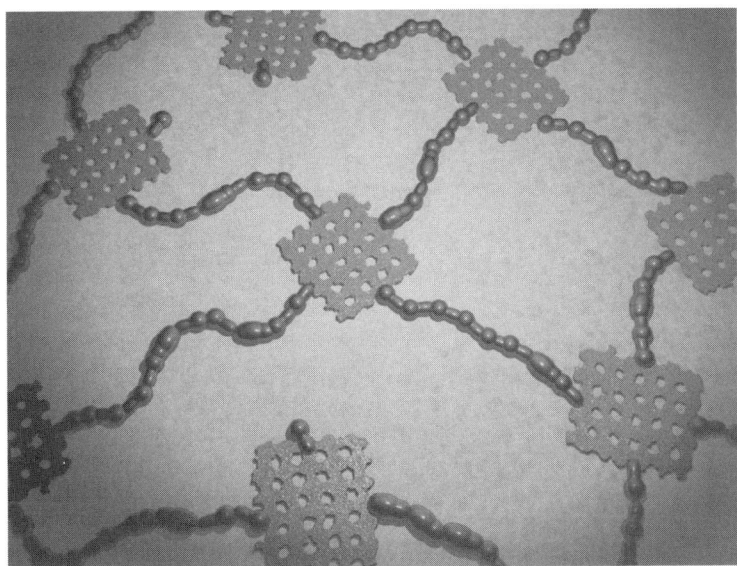

Figure 3.2 Schematic rendering of a self-assembled DNA interconnection network after metal decomposition.

The ensemble result demonstrates a new capability in designing self-assembled circuitry.

3.7.1 Overview

Figure 3.3 illustrates the design flow for the self-assembling process. The highlighted path through Figure 3.3 illustrates the sequence of deliverables (in italics) from the process. The design flow begins with an architectural description that is used to manually create a behavioral simulator that can verify the high-level procedural operations of the system. Once verified, the behavioral description can be used to manually capture gate-level modules for input to a sister transistor synthesis tool. These tools focus on circuit layout assuming a transistor-level module description is provided. Once a feasible layout is generated and verified it is used to create an ordered set of DNA sequence allocations for fabrication. The allocation process orders the assembly of the circuit so that a constant number of DNA sequences can be used independently of the circuit size.

3.7.2 Circuit Design

The circuit-level flow begins with a device (or transistor)-level description of the system generated by the synthesis tools; in this case, a suite of tools from

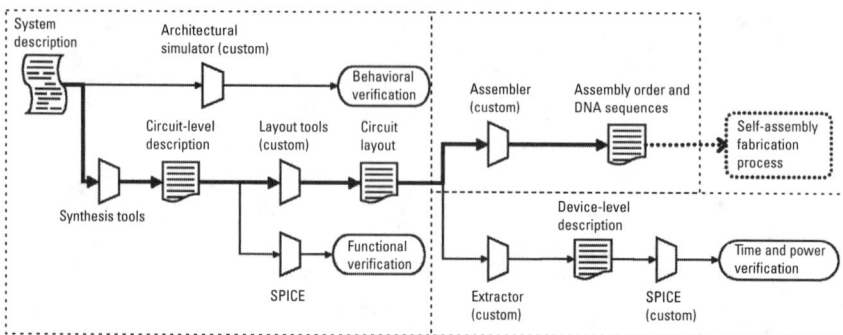

Figure 3.3 Design tool flow.

Mentor Graphics, Inc. produces this output. The transistor netlist is used to verify the functional properties of the circuit using a switch-level simulator. The unique constraints of the proposed self-assembling process (~100 × 100 FETs) make large circuits infeasible. Fortunately, the size of useful, but isolated, circuits rarely exceeds 10^4 FETs. This means that a high resolution SPICE simulation can be used to verify circuit functionality before the layout process. In this limited sense, the self-assembling technology simplifies the design flow compared to a conventional technology because it forces system architects to explicitly partition their designs into smaller, more easily factored logic blocks.

The layout process can include automated tools and/or manual full-custom steps. Custom layout tools provide an interface for manual layout and the exploration of design spaces through automated techniques. The constrained circuit sizes in this technology appear to make it more feasible to apply fully automated layout generation to the entire system than with conventional technologies.

The generated layout is used by a custom circuit extractor to back-annotate the original circuit with wire models and additional parasitics derived from the geometry of the layout. This requires the extractor to model the as-fabricated structure of the circuit and use geometric relationships to refine the circuit. The back-annotated circuit is simulated by the modified SPICE kernel with the empirical device models described in Section 3.7.3. The results of this simulation are used to verify the timing and power constraints of the circuit and can be used to make decisions that feedback to the system description and earlier design process. For example, if the verified circuit violates the power-delay product constraint, the next circuit can be adjusted accordingly.

3.7.3 Device Design

The device-level flow begins with the transistor level description of logic modules that have been synthesized and optimized by a logic design tool. This description is first verified using a switch-level simulator and then processed by an automated place and route layout generator. A SPICE deck is then extracted from the generated layout to estimate performance including wire delays and other parasitics. The results presented in this chapter use a modified SPICE 3f5 kernel similar to [14] and a custom semi-empirical device model for the CNFETs and parasitics [15, 16] to estimate the performance of the circuitry. The obtained results are consistent with the reference data used in the models and recent observations of CNFET performance [17, 18].

3.7.4 Self-Assembled DNA Design

In an analogous fashion to the generation of masks from layout artwork, DNA sequences are used by DNA strand manufacturers to create the DNA tags that will be attached to devices. These strands ultimately direct the assembly of the circuit during the fabrication process.

The layout is used by a custom assembler to render the layout into an ordered sequence of assembly steps and DNA tag identifiers. The assembler also generates the association of components (nanotubes or nanoparticles) with DNA tags.

The assembler uses the stitching algorithm illustrated in Figure 3.4 to order the DNA tag allocations. This boustrophedonic (i.e., literally as-the-ox-

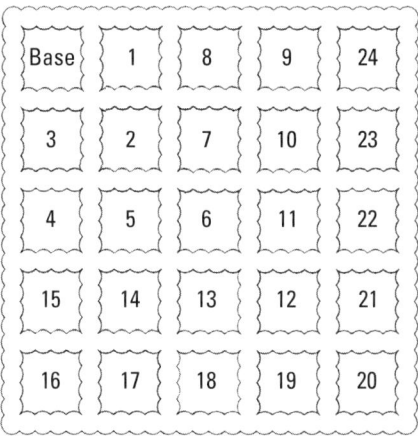

Figure 3.4 The assembly order follows a radial boustrophedonic pattern to build the grid.

goes) pattern is better than simple line scans because it minimizes the span of the grid at all stages.

The assembly ordering assumes a single "active" cavity (i.e., one available for binding nanotubes, etc.) in the scaffold at each step in the process. That is, it assumes the scaffold is extended in the direction of the next cavity (as specified by the assembly pattern) before each assembly step. To prevent components from binding to previously assembled cavities, the grid can be passivated with short DNA fragments that bind to unused locations on the cavity after each assembly step. At each step the assembler-generated set of tagged components, specified in the layout for that cavity, are added to the assembly environment. Figure 3.5 illustrates the positions of each tag in the cavity.

Each position has four associated orientations: north, east, south, and west. That is, a DNA tag can bind a component at any of its orientations. Further, nano*particles* bind to a reserved portion of the DNA tag and do not prevent nano*tubes* from binding to the same tag (i.e., nanotubes have their own portion as well). This distinction ensures that components can be selectively fused together to create circuits later.

A tag and orientation are specified for each attachment point of a component. For example, a nanoparticle can be specified with a single tag and orientation. The specifier N7 represents the north (side) of location 7. By design, a nanoparticle at N7 and a nanotube connecting (N7, S5) will fuse during the metallization process.

The exact sequence of the DNA used for each tag is taken from a pool of sequences that are known not to interfere with each other. The pool is generated using the thermodynamics-based methods described in Chapter 2.

The output from the assembler—the assembly ordering and DNA tag allocation—is then used to direct the self-assembly of the circuit. The next section describes the application of these tools to several simple logic structures.

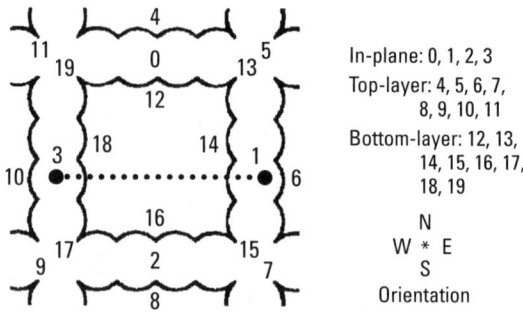

Figure 3.5   Tag location legend. Each number represents the location of a DNA tag. N, E, S, and W designate an orientation with respect to the location.

3.8 Case Studies

The design methodology is demonstrated by designing a NAND gate, full-adder, and SR-latch. While these circuits are trivial, the process introduces the tools needed for self-assembly. Each circuit layout was generated manually and converted to a SPICE netlist for switch-level verification. The NAND, and SR-latch layouts are illustrated in Figures 3.6 and 3.7, respectively. Table 3.1 lists several simple measures of the NAND, full-adder, and SR-latch layouts.

Each cavity has an area of 4×10^{-4} µm² determined by the spacing of the DNA scaffold, which has been designed to span a ~20-nm pitch. In terms of transistor density the cavity area corresponds to 2,500 transistors/µm² or approximately 20 times the density of current CMOS technologies (e.g., 45-nm CMOS has ~120 transistors/µm²).

Table 3.2 lists the transition energies (E_t) and switching delay (t_d) for each circuit simulated using the method described in Section 2.4. These results were obtained by loading each circuit output with an FO-4 inverter tree (i.e., four parallel inverters) and conditioning each square wave input through a series of four CNFET inverters. These testing conditions mimic typical circuit topologies by smoothing sharp input signals (i.e., sharp input signals can

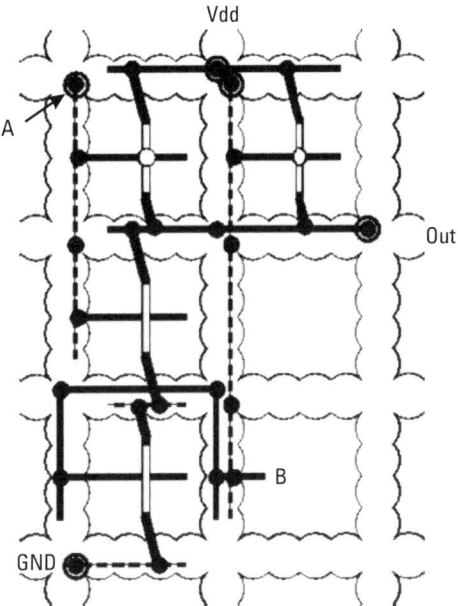

Figure 3.6 NAND gate layout (boundary cavities have been removed). Only nanotubes, particles, and scaffold DNA are shown.

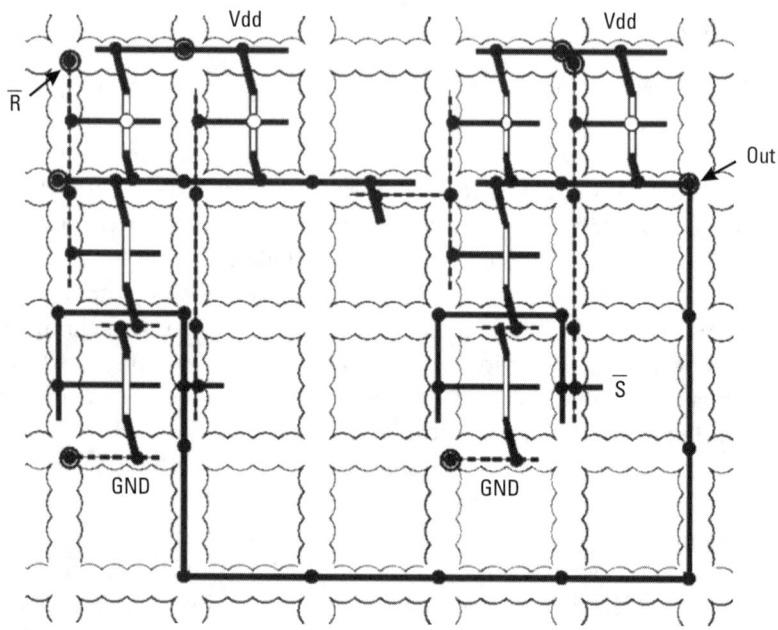

Figure 3.7 SR-latch layout.

Table 3.1
Layout Features

	Span (Cavities)	Layout Efficiency (FETs/cavity)
NAND	3 × 2	0.67
Full-adder	10 × 12	0.23
SR-latch	4 × 5	0.40

Table 3.2
Circuit Performance

Technology	Circuit	E_t (10^{-18} J)	t_d (ps)
CNFET	NAND	40	24
	Full-adder	192	104
	SR-latch	94	16
180-nm CMOS	NAND	19×10^3	60
	Full-adder	47×10^3	250
	SR-latch	28×10^3	140

make a circuit appear faster) and placing realistic loads on the outputs. The results in Table 3.2 are from the worst-case single-input transition event for each circuit. For comparison, simulation results from a 180-nm CMOS technology are included. The CMOS circuits here are identical to the CNFET circuits with the exception that CNFETs have been replaced with minimum size N-FET and P-FET transistors. The variation in the relative E_t and t_d between the circuits reflects poorly chosen transistor sizes (or multiplicity of CNFETs) and results in an imbalanced design. Even so, the performance improvement of CNFET logic over CMOS logic is evident.

Table 3.3 illustrates the assembly sequence of the NAND gate as generated by the tool. Each row represents a single active cavity (indicated by the number in braces as specified in Figure 3.4) and the assembly actions needed to complete it. For example, steps 0 through 2 are used to extend the grid in a particular direction to follow the stitching pattern, and steps 4 through 14 assemble the components for the cavity (cavity 2 is active). In this table, V indicates a Vdd connection (1.0V) and G indicates a ground connection (0V). These connections could be made out of the plane of the grid (or node) to microscale contacts above and below the grid (separated by a dielectric). This process will require assembly techniques or power-by-wire circuit designs described in Chapter 6.

Some cavities are not used in this layout due to the placement of the circuit on an overly large grid (in this case a 5 × 5 cavity grid). Table 3.3 illustrates how tangled the self-assembly process can become even for trivial design problems (e.g., a NAND gate). This underscores the need for the continued development of design tools capable of handling this emerging technology.

3.9 Conclusions

The development of self-assembling technologies that can either replace or supplement existing silicon technologies for fabricating circuits exposes many new challenges. The primitives for circuit design must be modified to deal with the requirements of new DNA-based self-assembly fabrication. This requires new design automation tools because of the distinctions between self-assembling and conventional photolithography processes. This fundamental change in technology motivates the development of design tools that address these differences.

This chapter presented the design tools for one such self-assembling circuit technology. The technology uses DNA to programmably self-assemble the circuit components and requires design tools that are unique to those used for silicon CMOS.

Table 3.3
DNA Sequence Allocations and Assembly Order for a NAND Gate

Grid Extension	CNFETs	Metallic CNTs	Nanoparticles
{2}	P-type:	5. (E3, W1),	1. G15, 3. V7,
0. base	7. (S0, N2)	13. (S19, N17),	6. E18, 8.N8,
2. East		14. (E11, W5)	9. S4, 10.W5,
4. South			11. S19, 12. N17,
{5}	N-type:	18. (E3, W1),	19. E18, 21. N16,
15. West	20. (S0, N2)	25. (S19, N17),	22. S4, 23. W5,
16. South		26. (E11, W5)	24. S19
17. East			
{6}		32. (S19, N17),	28. W5, 29. E11,
27. East		33. (E11, W5)	30. S19, 31. N17
{7}	P-type:	35. (E3, W1),	36. E18, 38. N8,
34. North	37. (S0, N2)	43. (S19, N17),	39. S4, 40. E11,
		44. (E11, W5)	41. S19, 42. N17
{8}			46. G17
45. North			
{10}			49. V9
47. East			
48. South			
{13}		55. (S11, N9),	53. S19,
50. South		56. (S19, N17)	54. S11
51. South			
52. West			
{14}	N-type:	58. (E3, W1),	59. E10, 61. N16,
57. West	60. (S0, N2)	67. (S11, N9),	62. S12, 63. W5,
		68. (E11, W5)	64. E11, 65. S11,
			66. G17
{17}		73. (E19, W13)	72. E19
69. West			
70. South			
71. East			

The number in braces indicates the active cavity as specified in Figure 3.5. Each step is numbered.

References

[1] Mirkin, C. A., et al., "A DNA-Based Method for Rationally Assembling Nanoparticles into Macroscopic Materials," *Nature*, Vol. 382, No. 6592, 1996, pp. 607–609.

[2] Aldaye, F. A., and H. F. Sleiman, "Dynamic DNA Templates for Discrete Gold Nanoparticle Assemblies: Control of Geometry, Modularity, Write/Erase and Structural Switching," *Journal of the American Chemical Society*, Vol. 129, No. 14, 2007, pp. 4130–4131.

[3] LaBean, T. H., et al., "Construction, Analysis, Ligation, and Self-Assembly of DNA Triple Crossover Complexes," *Journal of the American Chemistry Society*, Vol. 122, 2000, pp. 1848–1860.

[4] Winfree, E., et al., "Design and Self-Assembly of Two-Dimensional DNA Crystals," *Nature*, Vol. 394, 1998, pp. 539–544.

[5] Park, S. H., et al., "Finite-Size, Fully-Addressable DNA Tile Grids Formed by Hierarchical Assembly Procedures," *Angewandte Chemie*, Vol. 45, 2006, pp. 735–739.

[6] Dwyer, C., et al., "The Design and Fabrication of a Fully Addressable 8-tile DNA Grid," *Proc. Foundations of Nanoscience: Self-Assembled Architectures and Devices*, 2005, pp. 187–191.

[7] Pistol, C. and C. Dwyer, "Scalable, Low-Cost, Hierarchical Assembly of Programmable DNA Nanostructures," *Nanotechnology*, Vol. 18, 2007, pp. 125305–9.

[8] Fu, Q., C. Lu, and J. Liu, "Selective Coating of Single Wall Carbon Nanotubes with Thin SiO2 Layer," *Nano Letters*, Vol. 2, 2002, pp. 329–332.

[9] Liu, J., et al., "Fullerene Pipes," *Science*, Vol. 280, 1998, pp. 1253–1256.

[10] Keren, K., et al., "Sequence-Specific Molecular Lithography on Single DNA Molecules," *Science*, Vol. 297, 2002, pp. 72.

[11] Braun, E., et al., "DNA-Templated Assembly and Electrode Attachment of a Conducting Silver Wire," *Nature*, Vol. 391, 1998, pp. 775–778.

[12] Liu, D., et al., "DNA Nanotubes Self-Assembled from TX Tiles as Templates for Conductive Nanowires," *Proceedings of the National Academy of Science*, Vol. 101, 2004, pp. 717–722.

[13] Yan, H., et al., "DNA Templated Self-Assembly of Protein Arrays and Highly Conductive Nanowires," *Science*, Vol. 301, 2003, pp. 1882–1884.

[14] Dwyer, C., L. Vicci, and R. M. Taylor, "Performance Simulation of Nanoscale Silicon Rod Field-Effect Transistor Logic," *IEEE Transactions on Nanotechnology*, Vol. 2, 2003, pp. 69–74.

[15] Fuhrer, M. S., et al., "Crossed Nanotube Junctions," *Science*, Vol. 288, 2001, pp. 494–497.

[16] Burke, P. J., "An RF Circuit Model for Carbon Nanotubes," *IEEE TON*, Vol. 2, 2003, pp. 55–58.

[17]   Appenzeller, J., and D. J. Frank, "Frequency Dependent Characterization of Transport Properties in Carbon Nanotube Transistors," *Applied Physics Letters*, Vol. 84, 2004, pp. 1771–1773.

[18]   Rosenblatt, S., et al., "High Performance Electrolyte Gated Carbon Nanotube Transistors," *Nano Letters*, Vol. 2, 2002, pp. 869–872.

4

Architectural Implications of Self-Assembly

Technology change is fuel for architectural innovation. Evolutionary changes in CMOS have inspired research on several important topics including wire dominated designs, power dissipation, and fault tolerance. However, manufacturing defects, power density, process variability, transient faults, bulk silicon limits, rising test costs, and multibillion dollar fabrication facilities are some of the challenges facing the continued scaling of CMOS. While architectural modifications (e.g., multicore) can provide some short-term relief, the semiconductor industry recognizes the importance of these issues and the need to explore long-term alternatives to CMOS devices and fabrication techniques [1]. A revolutionary technology change, such as replacing CMOS, is a potentially disruptive event in the design of computing systems.

Emerging technologies for further miniaturization have capabilities and limitations that can significantly influence computer architecture, and these require reexamining or rebuilding abstractions originally tailored for CMOS. This chapter explores the architectural challenges introduced by emerging bottom-up fabrication of nanoelectronic circuits.

DNA-based self-assembly of nanoscale components using inexpensive laboratory equipment may achieve tera- to petascale integration. Although much of this technology is in its infancy (i.e., demonstrated in research lab experiments), by studying its potential uses for building computing systems, architects can gain a deeper understanding of the limitations and opportunities involved while providing important feedback to the scientists developing the new technologies.

The DNA self-assembly process described in the previous chapters presents several challenges that must be addressed when designing a computing system. The three primary aspects of the fabrication process are: small-scale control of placement and connectivity within a single node (Section 4.1), large-scale randomness in node placement and interconnects (Section 4.1.2), and high defect rates (Section 4.1.3). These three aspects significantly impact architectural decisions (Section 4.2), particularly since conventional architectures assume precise control at both the small and large scale. DNA self-assembly, however, also introduces some new opportunities by leveraging the programmability of DNA sequences to enable new computing paradigms.

4.1 Technology Implications

4.1.1 Small-Scale Control

The ability of DNA self-assembly to achieve only small-scale control impacts architectural decisions in several ways. Three of the most significant are: limited space, limited coordination, and limited communication.

Limited Space. A 150×150 node can have a maximum of 22,500 CNFETs; however, on-node interconnect will reduce efficiency since a node only has two levels of interconnect. Furthermore, a portion of each node must be allocated as a "pad" for the DNA interconnect to other nodes.

The limited node size presents a trade-off in node design. At one extreme, we could design just a single node type that contains both computation and storage capabilities. However, since storage and computation circuits must share the node, each may be severely limited in capability. Alternatively, we could design a few specialized node types, some devoted to computation and others to storage. Even when designing a specialized node, the limited space impacts architectural decisions. For example, large state machines are not an option within a node since there is insufficient space for state storage. Similarly, the number of bits available in a storage node may be limited, thus affecting an architecture's word size.

Limited Communication. Without large-scale control, there is limited communication among nodes. Each node has four neighbors and there is no long haul communication. Furthermore, the connections between nodes are limited to single wires. Although the degree of each node or the number of connections between neighbors could be increased, each connection occupies precious edge space. In contrast, conventional CMOS designs exploit multiple metal layers for long-haul communication and large-scale control to create multiwire connections between components.

Limited Coordination. Conventional CMOS designs rely on precise control during fabrication to create sophisticated circuits (e.g., 64-bit adder with carry look-ahead). For this technology, if the most sophisticated node is a full-adder, then it is unlikely that 64 such nodes can be coordinated to efficiently implement a 64-bit adder. Coordination among nodes is limited to immediate neighbors and it is difficult to configure a priori a group of nodes to operate in a coordinated manner. Clearly, coordination must be implemented at some level to create more functional components, but tight coupling (such as in a 64-bit adder) may not be achievable at this level.

4.1.2 Large-Scale Randomness

The self-assembly process provides excellent control at the small-scale, however it cannot achieve such control at large scales. The resulting randomness introduces some additional issues that architectures must address.

Random Node Placement. The self-assembly process does not guarantee where any particular node will lie in the final circuit. Each node simply attempts to connect to other nearby nodes. The architecture and machine organization must accommodate this arbitrary placement of functional blocks.

Random Node Orientation. Similar to the random node placement, the assembly process we envision does not provide control over node orientation. Any system design must tolerate arbitrary node orientations and cannot make a priori assumptions on orientation. For example, it is incorrect to assume that the "east" side of one node will connect to the "west" side of its adjacent node.

Random Node Connectivity. Connections between nodes are not guaranteed to succeed during self-assembly. Therefore, it is possible for any node to have between zero and four functioning connections to its neighbors. The architecture must not make any a priori assumptions about available connectivity. When combined with random orientation, it is possible for nodes to connect in a triangular shape rather than the 2×2 grid one would assume with nodes that have degree four.

4.1.3 High Defect Rates

An inherent aspect of any self-assembly process is defects. Fabrication defects can influence node functionality and connectivity. Some interconnect defects cause the above problems with connectivity. While some aspects of fabrication can reduce the likelihood of defects (e.g., purification steps or overdesign of DNA tags), there will always be a significant number of defects and any architecture using these technologies must tolerate them.

4.2 Architectural Challenges

The above discussion exposes several aspects of this fabrication technique for nanoscale circuits that must be addressed by any architecture and its corresponding implementations. In this subsection, we enumerate several important challenges to developing an appropriate architecture for this emerging technology. This list is not exhaustive, but rather highlights some important challenges.

Designing Nodes. The architect must decide what functionality to place in each node. Should there be homogeneous nodes or heterogeneous nodes? If heterogeneous, then what types of nodes? How does node design affect connectivity/communication with other nodes, and what primitives should be provided?

Utilizing Multiple Nodes. Since individual nodes do not contain sufficient computation and storage to perform much useful work in isolation, an architect must determine how to exploit multiple nodes. This must be achieved given the above limitations on coordination, communication, placement, orientation, and connectivity.

Routing with Limited Connectivity. Traditional routing techniques may not apply since there is limited space for the complexity of dynamic routing and there are insufficient guarantees on node placement and connectivity to use conventional static routing.

Developing an Execution Model. The execution model embodies the software-visible aspects of the architecture and can be influenced by implementation constraints or instruction set requirements. For the envisioned fabrication technique, the execution model must overcome the severe implementation constraints outlined above while enabling a reasonable instruction set.

Developing an Instruction Set. Programmable systems require an interface that enables software to specify operations. Typically this is achieved by the instruction set architecture (ISA). The ISA may be influenced by the underlying capabilities of the technology. Given the fabrication technique, the architect must design an appropriate ISA that supports the above execution model.

Developing a Memory System. Storage is a crucial component of most computing systems regardless of the execution model. The ability to store values for future use and to name and find particular values is a necessary aspect of most computing paradigms.

Interfacing to the Microscale. An important aspect of any nanoscale system is the interconnection to larger scale components (e.g., microscale). This connection is necessary for at least providing an I/O interface for communication with the outside world. It may be possible for the architecture to exploit this interface in other ways.

4.3 Opportunities for New Architectures

The conventional view of system design incorporates layers of abstractions including the software, architecture, logical implementation, and physical realization levels. System designers can use the interface between each of these levels to compartmentalize design tasks. For example, a software programmer does not need to know anything about the system's implementation or realization levels. Software written for a given architecture will operate correctly for multiple implementations of that architecture.

Despite the convenience of clean abstractions, technological trends are blurring the lines between design layers and are creating new interactions between previously unrelated layers. One example is virtual machines (VMs) such as VMWare and Transmeta which implement the application-software-visible architecture (virtual architecture) in VM software, allowing more flexibility in the hardware/software interface beneath the VM layer. Designers can change the VM-software-visible architecture (physical architecture) in response to implementation or technology constraints without affecting application software that was written for the virtual architecture.

Technological trends are forcing a closer coupling of architecture to implementation and even realization. One such trend is the increasing delay for long on-chip wires relative to other circuit delays, which are decreasing. New architecture designs explicitly consider this low-level implementation issue. For example, the designs of both the instruction-level distributed processor (ILDP) [2] architecture and the nonuniform cache architecture (NUCA) [3] minimize the number of long wires required in an implementation. Power is another issue that permeates all levels of system design, increasing the coupling between architecture, implementation, and realization [4].

Even more interaction between the hardware levels exists in emerging nanocomputing systems. Numerous programmable logic array-like computers have been proposed recently, including NanoFabrics [5] and array-based architectures [6]. The physical architectures of these systems are not separable from their crossbar-like physical realizations. Similarly, the active network architecture in NANA [7] (described in Chapter 7) and the SOSA [8] (described in Chapter 8) architecture are intimately tied to their physical realization. Because

NANA and SOSA are fabricated with DNA self-assembly, the interconnections of their computational nodes are random. This randomly interconnected computational substrate necessarily impacts the physical architecture.

Future technologies are likely to further increase the interactions between design layers. Programmable self-assembly is an emerging fabrication technology that must be considered in the higher layers of the computer system design. The characteristics of self-assembled systems require architects to explicitly consider the fabrication process. Programmable self-assembly offers an opportunity to perform computation during the fabrication process itself. Before self-assembly, computation was performed either prefabrication (during design as precomputation) or postfabrication (at compile-time or run-time.)

DNA self-assembly is the predominant programmable self-assembly technique due to DNA's rich interaction model. By encoding a computational problem's inputs with appropriately chosen base-pair sequences, the DNA will combine to form the solution. Researchers have used DNA computation to solve several computational problems, including the directed Hamiltonian path problem [9].

4.3.1 The Temporal Aspects of Computing

Computation can be performed at three different stages in the computer fabrication process:

- *Prefabrication:* Computation can be performed during the design process that precedes fabrication of the computer. For example, an ASIC contains components that the developers have preconfigured to quickly perform specific computations such as Fourier transforms.
- *Postfabrication:* The most traditional time for performing computation is after the computer has been fabricated. General-purpose processors are a good example of this temporal aspect.
- *At-fabrication:* Programmable self-assembly of computer systems provides a new temporal opportunity for computation. The self-assembly process itself performs useful computation during fabrication of a computer system.

Prefabrication Computation. Before a computer is fabricated, the developers can plan ahead to specify how the system will perform specific computations in the future. A simple example is the use of a lookup table (ROM) that stores solutions to computations the computer is expected to perform.

More complex examples are application-specific (or custom) integrated circuits (ASICs), field-programmable gate arrays (FPGAs), or digital signal processors (DSPs), which inherently compute portions of their application space before fabrication—that is, they incorporate application-specific optimizations. Defect-tolerant design—for example, systems like the Teramac that use redundancy [10, 11]—is also a form of prefabrication computation with the goal of avoiding fabrication-time errors. Even general-purpose computers have some amount of hardware that is tailored to perform specific computations—for example, floating-point arithmetic.

Prefabrication computation is well suited to many types of computations, since designers often know ahead of time what computations a computer is likely to perform, or at least those computations it will perform frequently. However, prefabrication computation is resource-intensive both in terms of design time and final hardware.

Prefabrication computation involves more complex designs, and this added complexity leads to longer design times and more costly fabrication. For example, adding hardware to perform floating-point division is a form of prefabrication computation that complicates the system design and requires more chip area compared to having software synthesize floating-point division from more primitive hardware instructions.

Postfabrication Computation. Postfabrication computation is the conventional execution of a sequence of instructions—either micro- or macrocode—at the moment of interest. Although the fabrication is considered from a hardware perspective, it is possible to perform an analogous exploration of software. For example, software design could be considered to be prefabrication, compilation to be at-fabrication, and run-time execution to be postfabrication. From the hardware perspective, both compile-time and run-time are postfabrication.

Forms of postfabrication computation include compilation, program transformations, static scheduling, and numerical precomputation. General-purpose processors such as the Intel Pentium 4 perform the bulk of their computation postfabrication. Even ASICs and FPGAs perform some postfabrication computation that complements the computation they perform prefabrication. For the purposes of this discussion, consider general reconfigurable computing to have two components: (1) a prefabrication step to create the mechanisms that enable reconfiguration; and (2) a postfabrication phase that can include a single configuration then execution or alternately can configure, execute, reconfigure, and execute.

Postfabrication computation has the most flexibility, and it is the most general-purpose form of computation. Nevertheless, the system design can constrain postfabrication computation. For example, graphics cards include specialized programmable processors tailored to operate on pixels or vertices (graphics processing units, or GPUs.) Although they can sometimes be used for more general-purpose computation [12], their performance on arbitrary programs is likely lower than on a general-purpose microprocessor.

One potential drawback of postfabrication computation is that execution time and power consumption are strongly dependent on intuition and knowledge of the computational problem (as opposed to the definition) and its implementation. For example, matrix multiplication has many different implementations that can yield dramatically different performance results based on cache memory behavior.

At-Fabrication Computation. Most current computers exploit pre- and postfabrication computation, but the advent of programmable self-assembly creates the opportunity to perform computation during the fabrication of a computer. This at-fabrication computation occurs by executing a well-defined rule set during self-assembly. The computation presolves a problem space—for example, a Hamiltonian path—in preparation for a simpler postfabrication computation. Current fabrication techniques lack these capabilities for meaningful problems. Instead, developers must specify solutions to the problem space at design time (prefabrication) or compute it postfabrication.

DNA computing is one example of at-fabrication computation, with the base-pairs determining the rule set. During fabrication (annealing), all possible paths are fabricated, and a post-fabrication step searches for and extracts the correct solution.

Another example system that exploits at-fabrication computation is the proposed oracle computer [13] (described in Chapter 5). In this system, each self-assembled node is the solution to a predefined computational problem. For example, an oracle for n-bit addition would self-assemble nodes that represent each of the possible n-bit additions the system can perform as well as their solutions. An oracle computer performs virtually all of its computation at-fabrication. The only postfabrication computation it performs is a lookup to find the appropriate solution node. An oracle is similar to an at-fabrication version of a prefabrication ROM lookup table. Different oracles can be developed for specific problems—for example, a Hamiltonian path. Chapter 5 provides further details on oracle computers.

In contrast to the oracle approach, the distributed array multiprocessor (DAMP) [13] (described in Chapter 6) relies more on postfabrication computation, using at-fabrication computation only to generate a large set of random numbers.

Programmable self-assembly enables at-fabrication computation on an unprecedented scale—for example, DNA computing. With its scaling capabilities, this form of computation can reduce algorithmic complexity by orders of magnitude, thus reducing execution time and energy consumption.

4.3.2 Extending the Temporal Spectrum

Most computers exploit or combine multiple points on the spectrum. Even general-purpose systems exploit arithmetic algorithms at design time to improve execution time—multiply unit, square root, ROM lookup table, and so on—and ASICs generally still have a postfabrication computation component.

Extending the range of the temporal spectrum to include at-fabrication computation introduces a new point at which computing can occur. It is now possible to couple at-fabrication and postfabrication computation as in the oracle computer. Exploring the trade-offs between these aspects of computation involves determining which computational tasks should be performed when. Architects will base these decisions on the relative cost—including design and fabrication time, performance, and power.

Similar to the recent shifts in architectural decisions that place greater significance on technology capabilities, future programmable self-assembled architectures will place greater significance on fabrication capabilities. The simple layered system design must change to preserve the advantages of abstractions while avoiding the introduction of rigid boundaries between the architecture, implementation, and realization/fabrication levels. Without these changes—illustrated in the progression from Figure 4.1(a) to Figure 4.1(c)—architectures may not be able to exploit the full capabilities of emerging fabrication techniques.

To exploit the ability to perform computation during the fabrication process, architects must partition computational tasks into three temporal aspects. Shifts can take place from prefabrication to at-fabrication or from postfabrication to at-fabrication.

The shift from prefabrication to at-fabrication is likely to occur due to at-fabrication's advantages in terms of scaling. For example, the choice between

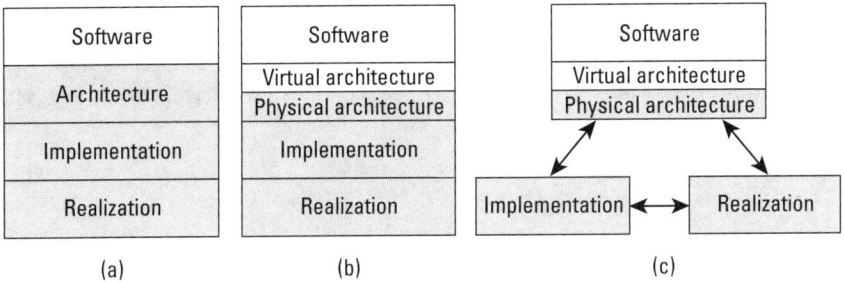

Figure 4.1 Layers of abstraction in system design: (a) conventional; (b) virtual machine; and (c) virtual machine coupled with technology. Shaded portions are hardware.

performing a computation using a prefabricated read-only memory design or at-fabrication creation may favor the oracle for larger problem sizes.

Programmable self-assembly's ability to efficiently preperform brute force computations will motivate the shift from postfabrication to at-fabrication. For example, the architecture could incorporate an oracle for a specific computation into a computer that is designed to solve that particular computation on a regular basis. The one-time cost of this at-fabrication computation would be amortized over the computer's postfabrication lifetime.

At-fabrication computation has an advantage over postfabrication computation for problems that require brute force because self-assembly has the potential to perform on the order of 10^{12} computations in parallel.

4.4 Summary

This chapter outlines the many challenges and opportunities created by a shift to a new fabrication technology. The challenge is to address each of these issues such that we arrive at a functioning system that can exploit opportunities. There are many possible approaches to developing DNA self-assembled computing systems. The following chapters present several case studies of architectures designed for use with DNA self-assembly.

References

[1] Semiconductor Industry, "International Technology Roadmap for Semiconductors," 2005.

[2] Kim, H.-S., and J. E. Smith, "An Instruction Set and Microarchitecture for Instruction Level Distributed Processing," *Proc. 29th Annual International Symposium on Computer Architecture*, 2002, pp. 71–81.

[3] Kim, C., D. Burger, and S. W. Keckler, "An Adaptive, Non-Uniform Cache Structure for Wire-Delay Dominated On-Chip Caches," *Proc. 10th International Conference on Architectural Support for Programming Languages and Operating Systems*, 2002, pp. 211–222.

[4] Mudge, T., "Power: A First-Class Design Constraint," *IEEE Computer*, Vol. 34, No. 4, 2001, pp. 52–57.

[5] Goldstein, S. C., and M. Budiu, "NanoFabrics: Spatial Computing Using Molecular Electronics," *Proc. 28th Annual International Symposium on Computer Architecture (ISCA)*, 2001, pp. 178–191.

[6] DeHon, A., "Array-Based Architecture for Fet-Based, Nanoscale Electronics," *IEEE Transactions on Nanotechnology*, Vol. 2, No. 1, 2003, pp. 23–32.

[7] Patwardhan, J. P., et al., "NANA: a Nano-Scale Active Network Architecture," *J. Emerg. Technol. Comput. Syst.*, Vol. 2, No. 1, 2006, pp. 1–30.

[8] Patwardhan, J. P., et al., "A Defect Tolerant Self-organizing Nanoscale SIMD Architecture," *Proc. 12th International Conference on Architectural Support for Programming Languages and Operating Systems*, 2006, pp. 241–251.

[9] Adleman, L. M., "Molecular Computation of Solutions to Combinatorial Problems," *Science*, Vol. 266, No. 5187, 1994, pp. 1021–1024.

[10] Heath, J. R., et al., "A Defect-Tolerant Computer Architecture: Opportunities for Nanotechnology," *Science*, Vol. 280, 1998, pp. 1716–1721.

[11] Culbertson, W. B., et al., "The Teramac Custom Computer: Extending the Limits with Defect Tolerance," 1996.

[12] Thompson, C., S. Hahn, and M. Oskin, "Using Modern Graphics Architectures for General-Purpose Computing: A Framework and Analysis," *Proc. 35th Annual International Symposium on Microarchitecture*, 2002, pp. 306–317.

[13] Dwyer, C., et al., "DNA Self-assembled Parallel Computer Architectures," *Nanotechnology*, Vol. 15, 2004, pp. 1688–1694.

5

Oracles and At-Fabrication Computation

5.1 Introduction

The previous chapter outlined the temporal space for computing. On the one end is DNA computing where the solution to a specific problem is determined through the hybridization of DNA and results are obtained using standard wet lab techniques. Oracles similarly exploit at-fabrication but also integrate standard electronics to enable limited postfabrication computation. In contrast to DNA computing, oracles are assembled using electrically active components (silicon rods, carbon nanotubes, and so forth) and DNA (metallized during postprocessing) to form large arrays of simple circuitry. However, oracles are designed to solve large portions of a particular problem during their fabrication (like DNA computing) rather than relying on purely postfabrication computation. That is, the oracles are a hybrid between the general-purpose electrical designs described in Chapter 3 and electrical circuitry inspired by DNA computing. This chapter describes the architecture of an oracle and provides several sample designs for solving specific problems.

5.2 System Overview

An oracle is designed to solve a problem space during its assembly. This architecture is enabled by DNA-guided self-assembly because DNA hybridization can enforce well-defined sets of pairing rules. So far, these rules have been used here to define the geometry of nanostructures to create circuitry; however, as in DNA computing, the rules can also be used to compute a result.

Abstractly, an oracle contains a large number of question and answer pairs. Questions are posed to it, and if the question is contained in any of the oracle's question/answer pairs, a response is generated. In this fashion, the oracle is similar to a large content-addressable memory (CAM) that has been preloaded with answers to certain problems. An oracle differs from a CAM by the method the question and answer pairs are entered into the array. To fully cover an input space of k bits, the CAM requires $O(2k)$ steps to load the answers (each of which must be computed in $F(k)$ time) for a total $O(2k) \cdot F(k)$, which scales with the problem complexity. Each address is an instance of the problem (or question) represented by up to k bits with its associated answer—just like a lookup table. In comparison, the oracle requires $O(k)$ steps to assemble and no run-time loading steps. The answers are determined by the rules that govern the oracle's assembly. The self-assembly of each question and answer pair provides the oracle with the answer (with a high probability, but not a certainty, that a given question and answer pair will exist within the oracle). If a particular question and answer pair did not form during the oracle's assembly, then the oracle cannot solve that instance of the problem and no response will be generated during such a query.

5.3 A Simple Oracle

A simple example of an oracle is the addition oracle (not useful in itself, but illustrative). The addition oracle has a simply defined problem and a compact functional description, and it performs all calculations at assembly.

Table 5.1 lists the entries in the truth table for a full-adder. The addition oracle will be assembled so that it incorporates many of the possible combinations of the truth table entries. Each line from the truth table represents a subproblem from an addition problem. A ripple-carry adder solves multibit addition problems by chaining the carry-out from one full-adder to the carry-in of the next full-adder. In a similar way the addition oracle will chain carry-outs to carry-ins, but at-fabrication rather than at run-time.

Table 5.1
Full-Adder Truth Table

A	B	Ci	S	Co
0	0	0	0	0
0	0	1	1	0
0	1	0	1	0
0	1	1	0	1
1	0	0	1	0
1	0	1	0	1
1	1	0	0	1
1	1	1	1	1

5.4 Implementation

Each line in the truth table is converted to a "tile" that represents a particular input and output combination, as in [1]. The difference in this work is that each "tile" is implemented by the self-assembly of nanoelectronic components and is electrically active. Many of the challenges in fabricating large networks from DNA will remain in fabricating the DNA scaffolded nanoelectronic circuitry considered here. However, progress is being made in this area [2] and discoveries in the fundamental properties of DNA self-assembly will apply to this theory directly.

The tile conversion method is similar to the way a carry-select adder speculatively precomputes a carry pattern and at run-time selects the proper carry path, with the exception that the oracle precomputes the entire carry path. The tiles that correspond to Table 5.1 are shown in Figure 5.1. Each tile has a carry input (on its upper-left), a carry output (on its lower-left), and an

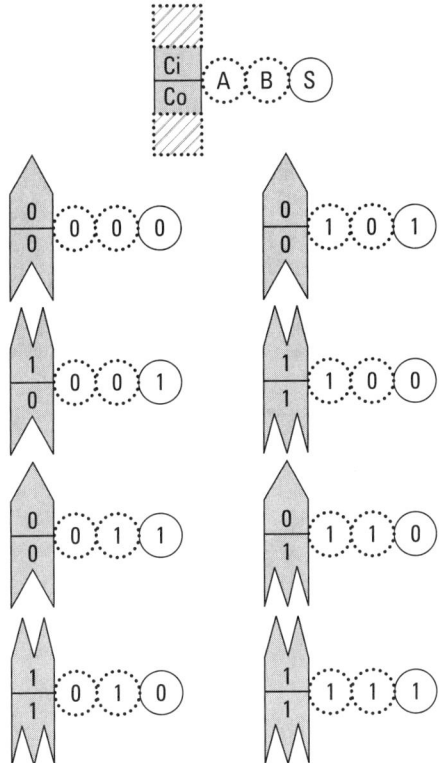

Figure 5.1 Assembly tiles for the addition oracle.

input-output triple (on its right). The carry bits are depicted in such a way that they fit together like the pieces of a jigsaw puzzle (i.e., 1s are double-dents or peaks and 0s are single-dents or peaks).

The iterative nature of the function (e.g., output from step i produces input for step $i+1$) allows strings of these tiles to implement an instance of the function's evaluation. For example, Figure 5.2 illustrates a simple 4-bit string made from the tiles in Figure 5.1. This particular example is an instance of the addition function for "3 + 5 = 8." The shape of the carries on each tile dictates how the string is formed. Valid strings must match each carry-out with the corresponding carry-in. In this fashion, the tiles perform an assembly-time computation as they form valid strings. The tiles only assemble into valid solutions for addition because the carries must match at each stage.

A complete addition oracle is the collection of all possible N-bit strings of tiles. Each string represents one particular input and output combination—for example, the "3 + 5 = 8" string shown in Figure 5.2. In this case, the string will respond with "8" to the question "what is 3 + 5?". For all other questions the string will be silent including the commutative question "what is 5 + 3?". The circuit complexity of each string is determined by the circuitry needed to read the string and respond to queries. A possible circuit for the addition oracle is shown in Figure 5.3.

The A, B, and S signals illustrated in Figure 5.3 carry the input query and response bits, respectively, for each tile. The OE_i and IE_i signals are the

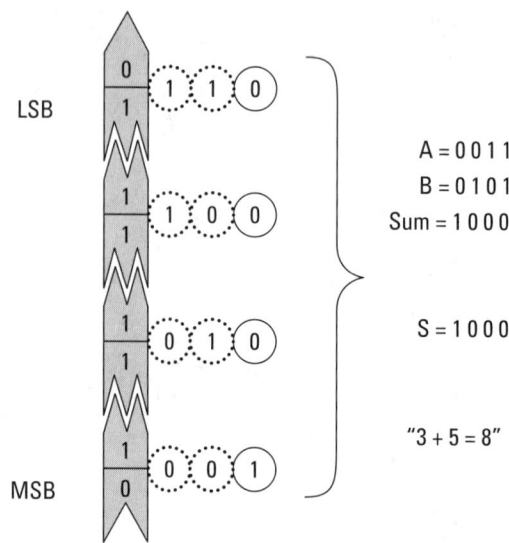

Figure 5.2 A 4-bit instance of the addition function. The carry-in and carry-out rules determine valid strings.

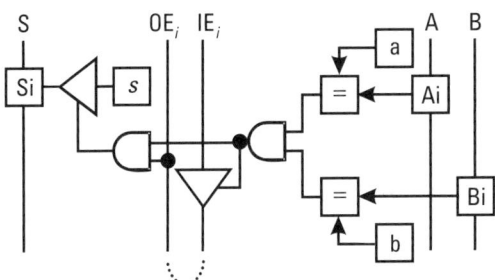

Figure 5.3 Circuit for an addition tile. The A, B, and S bits are constants assembled into a particular tile.

output and input enable signals, respectively, that coordinate the individual tiles in a string so that the string responds to a query if and only if all tiles in the string match the input query. The input enable signal is passed downward along the string and at the very last tile reflected upward as the output enable signal. Each tile can interrupt the input enable signal depending on the value of the current query, or the latched A_i and B_i input signals. Input queries are serially shifted into all strings (i.e., the circuits that implement each string) simultaneously. When the A and B values match the particular inputs of a string, all the tiles latch their sum values into the S_i latches. The only strings that respond to the query are those that have successfully reflected the input enable signal to their output enable line. The output enable signal can be used to trigger a ringer circuit that creates an oscillating signal that can be detected by an external receiver. This method is useful for problems that require only a single bit of output (e.g., NP-complete problems). Alternatively, the output enable signal can be used, as shown in Figure 5.3, to load the sum bit into a D-latch that can be shifted downward along the string to a ringer at the bottom that responds to the shift-out from the string.

5.5 At-Fabrication Computation Requirements

An example helps to visualize the resources required for at-fabrication computation, which is on the order of a few cubic meters of volume.

Assume the goal is to perform additions for 1,000 years on a 10-GHz processor. Given 31.5×10^6 seconds in a year, this yields 31.5×10^{16} additions per year, which is 3.15×10^{20} additions in a millennium. This must be mapped to some amount of work that can be performed with at-fabrication computation such as an oracle that computes the sum of two n-bit numbers at fabrication for all possible combinations of n-bit values (n^2, ignoring the commutativity of addition). The total computed additions can be used to determine what value n should be. The answer is the square root of (3.15×10^{20}), which is 1.77×10^{10}. Log_2 of this is approximately 34, and so it will take a millenium to compute all sums of all combinations of two 34-bit numbers operating at 10^{10} additions per second.

To estimate the material requirements to accomplish this with the DNA at-fabrication approach, assume that it is possible to assemble a single addition operation for two specific 34-bit numbers in a DNA structure (lattice) that is 800 nm on each side and 4-nm thick [3–5]. The volume for this is ($800 \times 10^{-9} \cdot 800 \times 10^{-9} \cdot 4 \times 10^{-9}$) ~ $= 2.56 \times 10^{-21}$ cubic meters.

To perform all additions, one lattice is needed for each combination of 34-bit values, or a total of 3.15×10^{20} lattices to perform 10^{10} additions. All the lattices stacked one on top of the other occupies about 0.8064 cubic meters or 806 liters (~200 gallons).

Of course, this is a lower bound on the volume of raw material. A larger volume is needed to allow for cooling, I/O, interconnect, power, and so on. Determining any other computation would require normalizing the work—and circuit volume—to that of the addition operation. This would either increase or decrease the volume based on what additional area is needed on each DNA lattice to implement the operation.

5.6 Generalization of the Oracle

The addition oracle serves as a simple model for other oracle designs. Carry chains are simple input and output constraints that can be generalized to include more complex relationships.

An oracle can solve a problem that can be expressed using the form illustrated in Figure 5.4. The functions F_i and G_i take the f_{i-1} and X_i inputs and generate the f_i and g_i outputs, respectively.

Each input (f_{i-1} and X_i) and output (f_i and g_i) are bit vectors. To aid in initializing the system, f_{-1} is assumed to be α, which is a constant defined at assembly-time.

Equations (5.1) through (5.3) describe the addition oracle using the form illustrated in Figure 5.4. These equations are derived from the truth table for addition, shown in Table 5.1. The input vector **X** has two elements, A and B, which represent the input operands. Equation (5.1) is the carry-out bit, and equation (5.2) is the sum bit.

$$F_i\left(f_{i-1}, X_i\right) = f_{i-1} \cdot \left(X_i[A] + X_i[B]\right) + \left(X_i[A] \cdot X_i[B]\right) \quad (5.1)$$

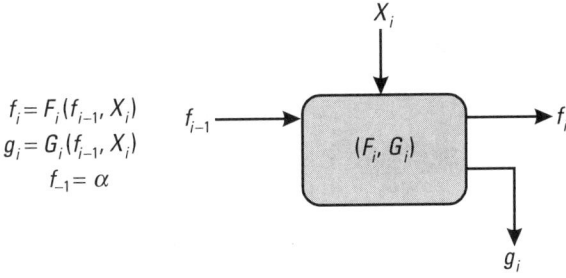

Figure 5.4 Problem expression solvable by an oracle.

$$G_i(f_{i-1}, X_i) = \overline{f_{i-1}} \cdot \left(\overline{X_i[A] + X_i[B]}\right) + \left(X_i[A] \cdot \overline{X_i[B]}\right) +$$
$$+ f_{i-1} \cdot \left(\overline{X_i[A]} \cdot \overline{X_i[B]} + X_i[A] \cdot X_i[B]\right) \quad (5.2)$$

$$\alpha = 0 \quad (5.3)$$

5.7 Hamiltonian Path Oracle

The Hamiltonian path (HAM-PATH) problem is NP-complete and represents what is considered to be an intractable problem. The problem consists of determining if a path exists through a connected graph of nodes such that the path visits each node exactly once.

The HAM-PATH oracle computes all paths through a fully connected graph at assembly-time in a manner very similar to the way Adleman solved the HAM-PATH problem using DNA [6]. The difference between the oracle and Adleman's approach is that the oracle solves the problem for any instance of the problem (with fewer than a fixed number of nodes).

Adleman's solution encodes each edge in a graph as a DNA fragment that has two sticky-ends representing the starting and ending nodes of the edge. Each node in the graph is allocated a sequence of DNA, and any edge that starts at that node will use this sequence on one end. The other sticky-end of the DNA fragment uses the complement of the DNA sequence assigned to the ending node. All of the fragments are mixed together and form strings of edges (in the form of DNA fragments) that represent feasible paths through the graph. Since Hamiltonian paths visit each node once, only strings with as many edges as there are nodes in the graph are feasible Hamiltonian

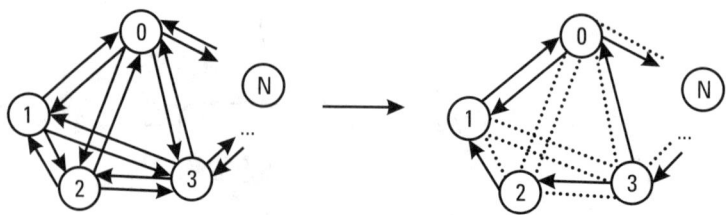

Figure 5.5 The fully connected graph on the left is collapsed to a particular graph on the right. The dashed lines represent deleted edges

paths. All other strings are discarded (using biochemical techniques). Cycles in the graph need special treatment [7]. The entire process takes on the order of weeks from start to finish.

The way the HAM-PATH oracle solves all instances of the problem is by solving the problem for a fully connected graph and then discarding solutions at run-time based upon a particular input graph. Paths from the fully connected graph that do not appear in the problem instance are deleted at run-time. This idea is illustrated in Figure 5.5.

Like the addition oracle, the HAM-PATH oracle uses random strings of tiles to perform an at-fabrication computation. The addition oracle formed all N-bit sums at assembly time. Likewise, the HAM-PATH oracle forms all paths through the fully connected graph. At run-time the HAM-PATH oracle selects the edges that exist in the current problem instance. After selecting the edges in the problem instance, one or more computing elements within the HAM-PATH oracle responds (electrically) to indicate that a Hamiltonian path exists through the graph if and only if it has a solution.

The design of the circuitry for each HAM-PATH tile is more complicated than the addition oracle tiles because they need to remove nodes from a set and then respond to selected graph edges. However, the generalization in Section 5.4 still holds and a more detailed account of this design and its circuitry can be found in [8]. The circuit level simulation results of this architecture estimate a run-time cycle period of ~10 μs per graph instance to compute the optimal path. This is approximately 40,000 times faster than the estimated Earth Simulator performance (200 ms).

5.8 Block Edit Oracle

The block edit problem is found in several application domains including molecular biology and functional genomics, pen-based computing, speech recognition, and natural-language processing. Briefly, the block edit problem involves selecting blocks of characters from one string (A), reordering the blocks, and then comparing them to characters from a second string (B). The optimal solution minimizes the number of edits required to transform each block of characters in A to the corresponding characters in B after the edit process. Variations of the problem specify whether the blocks must cover all characters from either or both strings and whether the transformation can be made with disjoint use of the blocks or not.

Some variations of the problem such as the noncovering and nondisjoint versions can be solved in polynomial time. Unfortunately, many of the interesting varieties of the block edit problem are NP-hard [9].

An NP-hard block edit problem can be solved optimally for a fixed input size—larger than is feasible with present day computers—using postfabrication and at-fabrication time computation. Although approximation algorithms may exist for a given problem, the advantage for oracles is in finding optimal solutions. Optimal solutions are required when the cost for a suboptimal solution is so great that the application can not be justified otherwise. For example, problems that involve the loss of human life or combat situations that involve considerable numbers of personnel and expensive equipment often require optimal solutions given the high stakes nature of the situation. The examples here present a mix of the temporal aspects of computing to achieve a range of postfabrication performance at the expense of increased manufacturing costs.

Table 5.2 summarizes three systems for solving the block edit problem, distinguishing between these systems based on how they divide the computational work between at-fabrication and postfabrication.

Table 5.2
Comparison of Systems for Solving the Block Edit Problem

System	Fabrication Material	Postfabrication Complexity
All postfabrication	None	$O[(N_A!)^2]$
At-fabrication block partitioning	$O(N_A!)$	$N_A! \cdot N_A \cdot N_B \approx O(N_A!)$
At-fabrication block partitioning plus at-fabrication block reordering	$O[(N_A!)^2]$	$O(N_A \cdot N_B)$

5.9 Purely Postfabrication Computation

A purely postfabrication optimal solution to the NP-complete block edit problem—cover and/or disjoint to cover and/or disjoint—naïvely requires $O[(N_A!)^2]$ running time, where N_A is the length of the input string. Figure 5.6 shows the brute force approach.

This method holds string B fixed while searching for a transformation of string A to string B. Since there could be as many as N_A single-character blocks, there are $O(N_A!)$ ways to partition string A. Each partition of blocks must be evaluated under all block arrangements, of which there are $O(N_A!)$, since in the worst case there could be N_A blocks.

The final step is to perform a per-block edit transform for each block partition and arrangement, which is $O(N_A \times N_B)$ running time with the use of a linear program. Combining these terms shows that in the worst case the brute force approach takes $O[(N_A!)^2 \times N_A \times N_B]$ or just $O[(N_A!)^2]$ to find an optimal block edit transformation from string A to string B.

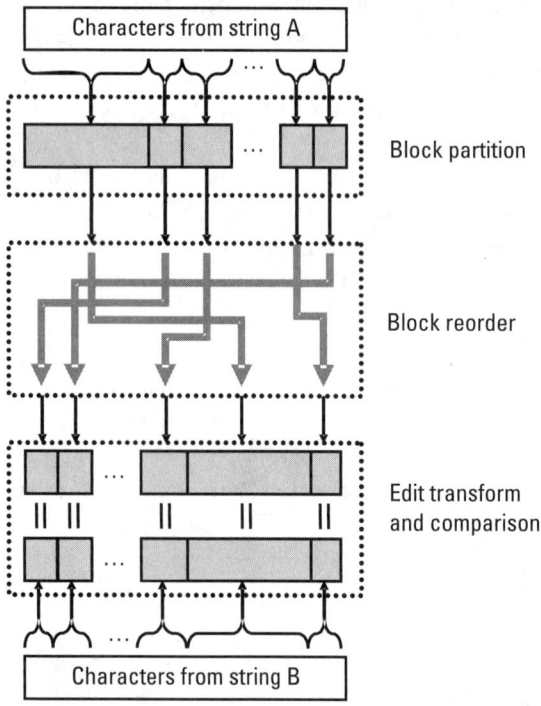

Figure 5.6 Overview of the brute force solution to the block edit problem. This solution holds string B fixed while searching for a transformation of string A to string B.

The astounding complexity of this problem means that, for a simple 15-word sentence (replacing characters with words maintains the constraint that the transformations do not destroy the sentence's lexicon), in the worst case it would take the better part of a millennium to find the optimal transformation to another sentence if each step takes only 10 femtoseconds (100-THz evaluation rate).

5.10 At-Fabrication Components of Optimal Block Edit Solutions

There are many alternatives for exploiting at-fabrication computation to solve the block edit problem. The only way to optimally solve the block edit problem is to perform an exhaustive search over the block partition space and the block reordering space. Here, the focus is on at-fabrication computation of block partitions and block reordering. The trade-off that architects must reason about is between the amount of material required at-fabrication and the amount of postfabrication computation.

5.10.1 At-Fabrication Block Partitioning

Block partitioning is the process of dividing the input string (A) into all possible substrings, where each substring is called a block. At-fabrication block partitioning is supported by using tiles, where each tile is the realization of one specific block. A tile can realize the block by either fabricating the appropriate substring directly during self-assembly, or by incorporating electronics that identify and store the appropriate substring when the string A is first input to the system.

Figure 5.7 illustrates a few tiles from the $O(N_A^2)$ set of tiles that the computation can use to partition the source string (A). Each tile represents a start and end position in A. The figure shows the tiles used to create all substrings that start with the first character in A up to the fourth character in A.

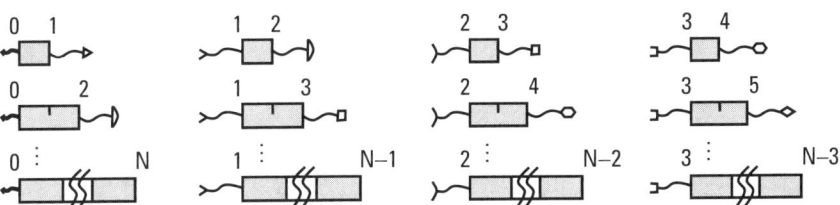

Figure 5.7 Sample tiles for creating substrings. The shapes on the tile ends represent the concatenation of the tiles to form a string partition.

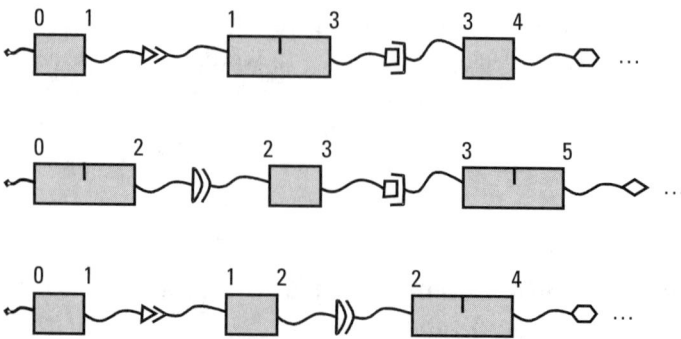

Figure 5.8 Block partition computations. The blocks from Figure 5.7 self-assemble randomly to form partitions. Each partition represents one of the $O(N_A^2)$ possible configurations.

A computation forms a string partition by concatenating the tiles in the original string. To achieve this, the computation specifies self-assembly rules such that a tile (T_1) will only bind to another tile (T_2) if the start of T_1 matches the end ordinal of T_2. This is represented in Figure 5.7 by the shapes on the tile ends.

Figure 5.8 illustrates some example partitions using the tiles from Figure 5.7. The first partition breaks string A into length 1, 2, 1, ..., blocks, the second into length 2, 1, 2, ..., blocks, and the third into length 1, 1, 2, ..., blocks. These partitions are randomly distributed over many such tile assemblies (formed at-fabrication time) to probabilistically cover the partition space. In this way, tile assembly computes partitions before any postfabrication computation, using $O[(N_A)!]$ material.

Given the partition assemblies that the computation creates, the remainder of the algorithm can be performed either in postfabrication computation or by using more at-fabrication computation. For the first option, the postfabrication complexity remains $O(N_A! \times N_A \times N_B)$ or just $O(N_A!)$. Shifting more of the computation to at-fabrication further reduces run-time complexity but at the expense of more material.

5.10.2 At-Fabrication Block Reordering

Figure 5.9 illustrates a block edit problem using at-fabrication block reordering. The five intermediate stages in Figure 5.9 perform all possible block swaps through self-assembly to probabilistically cover all possible block orders. The figure shows one block reordering from the complete set of all possible reorderings that the computation will construct. At each stage, a block

Oracles and At-Fabrication Computation 75

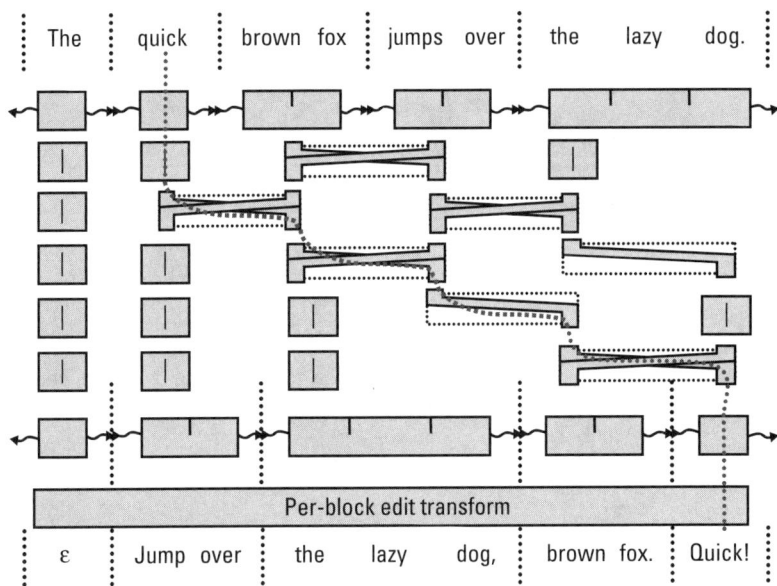

Figure 5.9 Block reordering. In the example, the word "quick" is reordered in each stage to arrive at its final position.

can remain in its current position, swap positions with an adjacent block, or shift over one position. These operations are encoded in tiles, and reordering occurs through tile assembly.

The highlighted path in Figure 5.9 shows how the word "quick" is reordered in each stage to arrive at its final position. In this way, the tiles can preselect a partition and block order to trade fabrication material for the $O[(N_A!)^2]$ run-time complexity of the brute force block edit solution. For this approach, the fabrication material required is $O[(N_A!)^2]$ and the postfabrication computation complexity is $O(N_A \times N_B)$.

5.10.3 Alternative Points in the Temporal Design Space

Two additional points in the design space help to show the range of the temporal computing spectrum: at-fabrication character edits and at-fabrication construction of all transformations and string instances.

The first alternative is an incremental approach that moves the next step in the algorithm (character edits) from postfabrication to at-fabrication. Even though well-known solutions exist to solve this problem in $O(N_A^2)$ running time, a simple solution exists at-fabrication. Figure 5.10 illustrates how a random sequence of insert and delete operations can transform the source block

Introduction to DNA Self-Assembled Computer Design

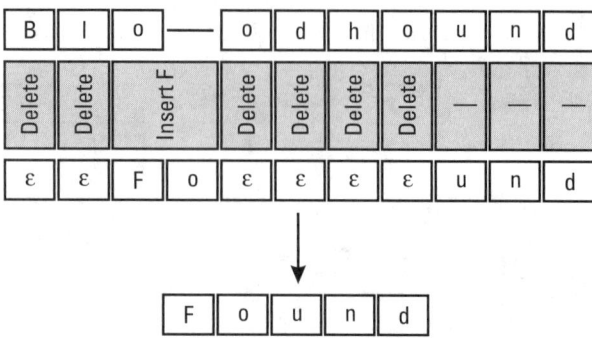

Figure 5.10 Self-assembled character edit sequence. In the example, a sequence of deletions and insertions transforms the source block "Bloodhound" to the target block "Found."

to the target block. A sequence of deletions and insertions transforms the source block "Bloodhound" to the target block "Found." The insert and delete operations self-assemble similarly to the block partition elements through a set of binding rules.

An even more extreme form of at-fabrication computation takes the method to the limit by fabricating not only block partition, ordering, and edit assemblies, but also every possible instance of the problem. For example, given a small enough length of strings, self-assembly can form all possible strings that are incorporated as inputs to the at-fabrication computations. This method is inefficient because the postfabrication task of communicating the two strings that define the problem instance to be solved can take as long as the alternative postfabrication solutions. This implies that a design should balance the ratio of postfabrication computation complexity to at-fabrication time material costs. DNA self-assembly enables this kind of computation because it can implement at-fabrication time computation *and* is low cost.

The primary distinction between the three systems for solving the block edit problem is how they divide the computational work between at-fabrication and postfabrication. Self-assembly provides new opportunities and design points for architects across the temporal spectrum of computing. Inherent in this is the trade-off between the material required for at-fabrication and postfabrication run-time complexity.

Programmable self-assembly provides new options for solving difficult computer architecture design problems. In particular, architects can use self-assembly to explore the engineering trade-offs involved in combining at-fabrication and postfabrication computation. Although at-fabrication computation is resource intensive, architects can use it to precompute solutions to problems that are too computationally intensive for postfabrication solutions. A spectrum of designs for solving the block edit problem illustrates the numerous possible design points that programmable self-assembly enables and the potential for novel architectures that exploit this capability.

5.11 Summary

Oracles represent an interesting point in the design space of self-assembled computer systems. They exploit the programmability of DNA to solve portions of a large design space during fabrication and then rely on electronic circuitry to help obtain the result quickly at run-time. Unfortunately, each oracle can solve only the problem it was designed for. The remaining chapters discuss architectures that move further along the temporal computing spectrum toward postfabrication execution, and thus provide increased general purpose computing.

References

[1] Yan, H., et al., "Parallel Molecular Computations of Pairwise Exclusive-Or (XOR) Using DNA 'String Tile' Self-Assembly," *Journal of the American Chemical Society (Communication)*, Vol. 125, No. 47, 2003, pp. 14246–14247.

[2] Sa-Ardyen, P., N. Jonoska, and N. C. Seeman, "Self-Assembly of Irregular Graphs Whose Edges Are DNA Helix Axes," *Journal of the American Chemical Society*, Vol. 126, 2004, pp. 6648–6657.

[3] Yan, H., et al., "DNA Templated Self-Assembly of Protein Arrays and Highly Conductive Nanowires," *Science*, Vol. 301, 2003, pp. 1882–1884.

[4] Patwardhan, J. P., et al., "Circuit and System Architecture for DNA-Guided Self-Assembly of Nanoelectronics," *Proc. Foundations of Nanoscience: Self-Assembled Architectures and Devices*, pp. 344–358.

[5] Dwyer, C., et al., "Design Tools for a DNA-Guided Self-Assembling Carbon Nanotube Technology," *Nanotechnology*, Vol. 15, 2004, pp. 1240–1245.

[6] Adleman, L. M., "Molecular Computation of Solutions to Combinatorial Problems," *Science*, Vol. 266, No. 5187, 1994, pp. 1021–1024.

[7] Bourianoff, G., "The Future of Nanocomputing," *IEEE Computer*, Vol. 36, 2003, pp. 44–53.

[8] Dwyer, C., *Self-Assembled Computer Architecture: Design and Fabrication Theory*, Ph.D. dissertation, Chapel Hill: University of North Carolina, 2003.

[9] Lopresti, D., and A. Tomkins, "Block Edit Models for Approximate String Matching," *Theoretical Computer Science*, Vol. 181, No. 1, 1997, pp. 159–179.

6
The Distributed Array Multiprocessor

6.1 Introduction

The early limitations of a self-assembling realization technology require small circuits. Single bit, serial processing elements (PEs) are well suited to such limitations. Bit-serial processors require less circuitry and have simple interfacing requirements. This can be placed into the context of the temporal aspects of computing as described in Chapter 5. The trade-off between assembly-time and run-time complexity has at one end the oracle, described in Chapter 5, and at the other end traditional sequential and parallel machines. The oracle has nearly no run-time complexity because its computation is performed during the assembly process. This limits the oracle to solve one particular class of problem, but in its entirety. Greater flexibility comes by introducing more run-time capabilities into the machine design. Thus, the goal of general-purpose computing requires machine designs that oppose the oracle on this spectrum.

The distributed array multiprocessor (DAMP) employs much greater run-time computation to achieve greater program flexibility than the oracle. In this sense, the DAMP is more practical because it has broader applicability that can offset the cost of developing the self-assembly techniques required to build the machine.

6.2 System Overview

The DAMP is similar to traditional single-instruction multiple-data (SIMD) machines with two important differences: it has no inter-processor communication, and many, many more processors. The most significant difference is the lack of any communication hardware between processors. The large machines found in supercomputing centers today have high-bandwidth interconnections between processors. The processors in the DAMP have no way to communicate with each other (i.e., they are decoupled, or distributed, from each other) except through a shared control unit. This limited form of

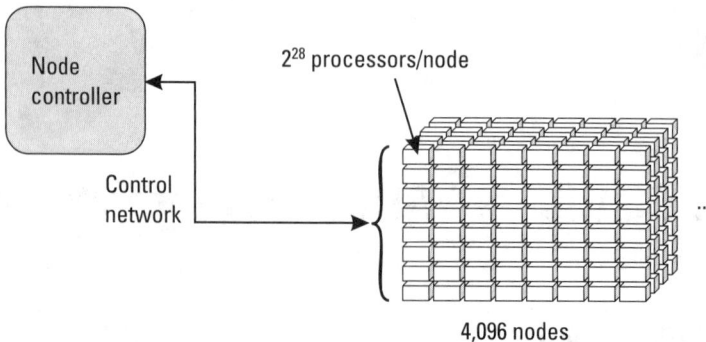

Figure 6.1 The node controller and processor node arrangement.

inter-processor communication limits the DAMP to embarrassingly parallel problems that involve little to no data sharing.

The magnitude of the number of processors in each machine type is also dramatically different than typical SIMD machines. Self-assembly promises on the order of 10^{12} processing elements that far exceeds typical processor counts. However, the complexity of any individual PE is greatly diminished with respect to the processors used in traditional SIMD machines. The basic structure of the DAMP is illustrated in Figure 6.1. The node controller sends control signals to each processor node in parallel. Each processor node can detect a PE-generated signal from an individual "ringer" circuit embedded in each PE. This reduced output capacity (from the processor's perspective) is a conservative worst-case scenario to reduce the fabrication complexity of each PE.

For comparison, an Intel Pentium 4 (P4) has ~100 × 10^6 transistors, while a single DAMP processing element has ~1,600 transistors. That is, a single P4 has the equivalent of 62,500 DAMP processing elements. However, the entire DAMP has ~1.75 × 10^{15} transistors, or the equivalent of ~17,500,000 Intel Pentium 4 processors. A single P4 (and its accompanying hardware) occupies about a 0.08-m³ volume; the number of P4s that are equivalent to the entire DAMP (12-m³ volume) would completely occupy 2.5 × 10^6 cubic meters or a square room ~3,300 feet on a side with a 6-foot high ceiling. Moreover, the self-assembly process may have limited the complexity of processing elements but the linear and volumetric scaling of the devices present a clear advantage over conventional photolithographic processing.

6.3 Execution Model

Figure 6.2 illustrates the basic execution model for a DAMP processing element. Each PE has five 16-bit registers and conditionally executes the global instruction stream depending on the value of its wait-status bit. In this bit-serial design, the least significant bit is the first bit to participate in each operation; the bit at the bottom of a register in Figure 6.2. Through separate shift controls the accumulator can shift independently from the R0–R4 registers enabling relative data shifts. The operational unit is a full-adder that can provide either the carry out or sum signal to the accumulator input. Each register R0–R4 can receive either its own LSB or the accumulator output as input during a shift. The 6 status bits can be used to implement a wide range of conditional operations (see Section 6.5.1).

The 16-bit accumulator, R0, and R1 have the ability to load a random constant that is probabilistically unique to each PE through a limited form of at-fabrication computation. The constant relies on a random assembly event to select either a 1 or 0 for each bit in the register (e.g., by a conductor or insulator attaching randomly to a latch's load line). The random constant is used as an index or a seed for selecting a portion of a problem space. At runtime a program can supplement these at-fabrication input bits with a counter or value from the node controller in the least significant bit positions. This random constant is fundamentally uniform and simply the starting point for an application-specific randomization procedure.

The DAMP processor (i.e., the ensemble collection of hardware) has $\sim 10^{12}$ processing elements that can probabilistically evaluate any 40-bit input

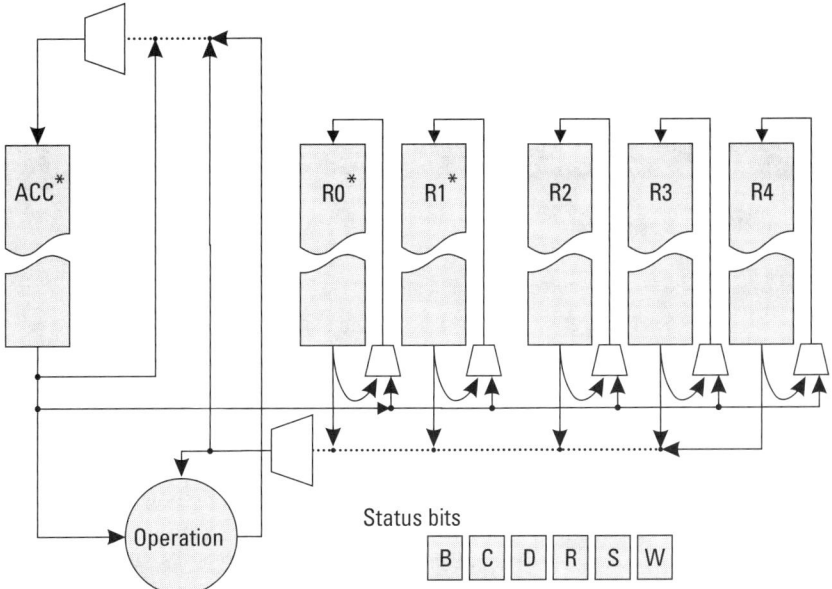

Figure 6.2 Processing element diagram. ACC, R0, and R1 can be loaded with random bits.

space with only one run of a global program. The program instructs each PE to manipulate its random constant to produce an answer to a problem. If the answer is satisfactory (e.g., below some threshold, etc.) the PE with the answer can alert the node controller through a shared medium. A binary search ensues over the input space (i.e., the random constants) to determine all the bits from the random constant used by the successful PE. The time-complexity of this search scales with the number of unique solutions since each solution requires a distinct branch during the binary search.

The self-assembly fabrication technology used to build the DAMP demands that each processor be simple. This drives the use of bit-serial processing elements and a simple controller. As a consequence all instructions are software encoded without the need for microcode (e.g., as in VLIW machines). A more complete list (than will fit here) of instructions that can be efficiently implemented by the DAMP, including the gate level implementation details, transistor level nanorod layout for the processing elements, and the behavioral simulation details can be found in [1, 2].

Each DAMP processing element uses a set of status bits to coordinate the activity of instructions during execution. The six status bits can be used to implement a wide range of operations. The definition and operation of each status bit is listed in Table 6.1.

The random constants in each PE are a peculiar feature of this architecture which enables the DAMP to tackle large combinatorial problems. The constant replaces the processor index commonly used by SIMD machines. In a manner similar to DNA computing [3, 4] the random constants can be cast

Table 6.1
Description of DAMP Status Bits

Status Bit	Description
B	Set to the current operand bit from R0–R4
C	Set to the current carry out from the current operation
D	Detects a one on the output of the current ALU operation
R	Ringer control (described in Section 6.4.3). The controller can detect when any of the processors have set R = 1
S	Set to the current sum output
W	The wait bit. Set according to the value of another status bit (B, C, D, or S), or cleared. The machine ignores instructions when W = 1 with the exception that the RESUME instruction can clear W (W = 0) and start normal instruction execution

Figure 6.3 DAMP input space.

as instances of problem variables. The input space for the DAMP is illustrated in Figure 6.3.

The 2^{40} processors in the DAMP suffice to probabilistically evaluate any 40 bit input space with only one run. If the input space of a problem cannot be fit into 40 bits, additional bits computed at run-time may be used to augment them. By using a counter as the least significant bits of the input space, the 40 random bits can be treated as the most significant bits of the input space. After each run of the program the counter is incremented. This enables the DAMP to uniformly cover the larger input space provided that the random bits uniformly cover the 40-bit input space.

6.4 Hardware Design

Connecting self-assembled structures to power and I/O electrodes could easily be a manufacturing bottleneck given the vast number of structures that can be assembled at once. This section describes two interconnection methods for large numbers of devices (10^{12}) that can be manufactured on large scales.

6.4.1 Modular Assembly

The placement of each self-assembled processor must proceed unambiguously. One simple way to assemble mesoscopic-scale objects unambiguously is to rely on geometric features and surface energy minimization [5, 6]. This method of self-assembly works by creating modules with a particular geometric shape that can fit into a hole on the surface like a key. Materials coated on the sides of the module act like glue to keep the module in place. The module shape, or footprint, should be rotationally and reflectionally asymmetric so that a well-defined module orientation on the surface can be maintained.

Adopting a standard footprint among modules facilitates the mesoscopic-scale assembly if that footprint allows only a single final resting-orientation. This technique is used in fluidic self-assembly where an object minimizes any fluid shear force when it lands on the substrate in a strictly unambiguous fashion.

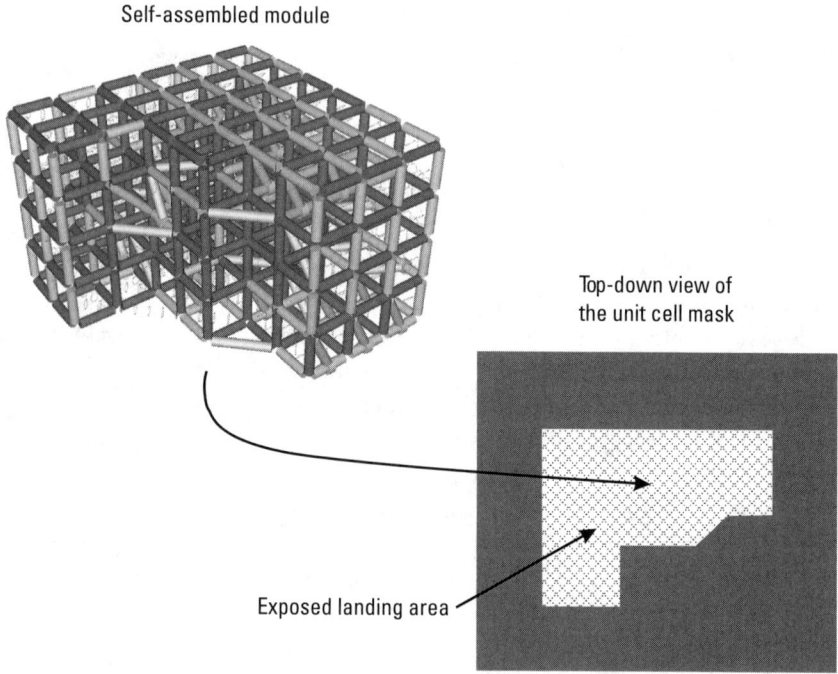

Figure 6.4 Modular assembly unit cell mask and module. The module lands in the exposed landing area. A new photoresist mask is constructed and the process repeats.

Lithographically patterned planar substrates, as illustrated in Figure 6.4 could be used as landing areas for the modules if the exposed and developed portions of the resist layer create recesses that fit the outlined shape of the self-assembled modules. Maintaining a registered substrate-to-mask alignment will allow multiple self-assembled modules to be stacked on top of each other [7]. The tolerances for the lithography process need to be within the range of the device feature size.

Modular assembly simplifies the interconnection problem by disambiguating the orientation of structures as they land on the substrate. Since the lithography pattern prevents improperly oriented modules from landing, the final structure will have a well-defined shape and orientation, as illustrated in Figure 6.5. The particular photoresist chosen to form the cavities must support high aspect ratio features and have a low surface energy. The commonly used SU-8 photoresist has high aspect ratio features but must be doped to reduce its surface energy. Research in this area has shown that it is possible to formulate epoxy-based photoresists and that the siloxane nature of the polymers makes self-assembled monolayer treatments feasible [8, 9].

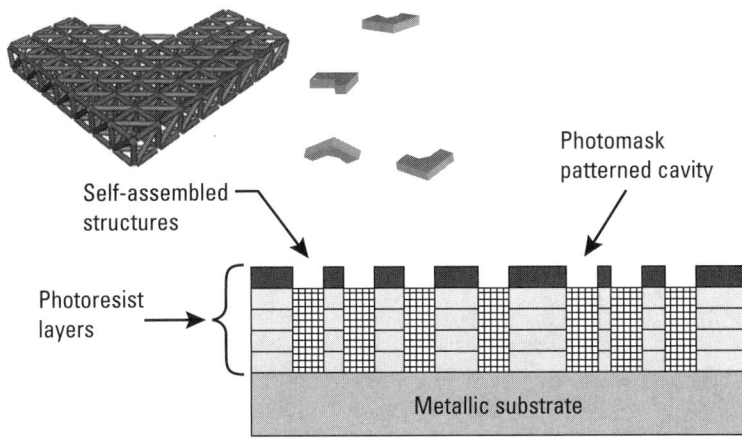

Figure 6.5 A modular assembly of stacked structures. Photolithographic steps create the sidewalls that corral the asymmetric structures.

The asymmetry in the footprint of each structure makes it possible to enforce a role for the substrate electrode and the top electrode (deposited during post-processing). However, a lithographically prepared silicon substrate (instead of a metallic substrate) can be used with conventional VLSI techniques to make a footprint capable of PE communication. The circuitry that controls the communication to the processors can lie beneath the footprints and be shared by several stacks.

6.4.2 Monolithic Assembly

Another solution to the I/O problem is to use fully self-assembled structures sandwiched between two power electrodes. This method is applicable to structures that are fully assembled before finishing the interconnection method. The most serious drawback of this method is power consumption due to the large capacitance of the sandwich. However, energy recovery through inductive loads may be possible if a high frequency AC signaling method to PEs can be used.

Figure 6.6 illustrates the layered interconnect method. Each electrode serves a dual purpose. The bottom electrode (P0) is used to electrically ground the circuitry and is also used as a clock signal. The top electrode (P1) is used to supply a positive voltage to the circuitry and is also used as a data signal.

This arrangement requires special power-up circuitry if modular assembly is not used to orient the structure as to which direction is up (the positive voltage electrode). This same circuitry, through the use of a bridge rectifier,

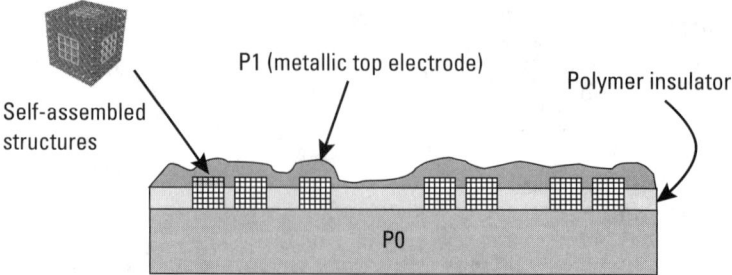

Figure 6.6 Layered interconnect method. The bottom electrode serves as ground while the top electrode serves as V_{dd}.

can supply power and provide a reference for how to interpret each electrode. By alternating between power and signaling phases, the electrodes can be used for both purposes. The initialization routine for this system follows:

The positive electrode (P1 in Figure 6.7) is slowly ramped to V_{dd} (1V in the circuits considered here) while the ground electrode (P0 in Figure 6.11) is connected to the system ground (0V). After some time, the orientation capacitors (C_{or0} and C_{or1}) will have fully charged or failed to charge depending on which electrode was powered up and which was grounded. At this point the power-up circuit can define which electrode (P0 or P1) is the positive (and data) electrode and which is the ground (and clock) electrode. The signals F0 and F1 will reflect this orientation and select the proper electrode to be connected to the internal DATA and CLOCK wiring.

To signal a logical 1, the positive electrode and ground electrodes are temporarily held high. The ground electrode is returned to ground after a fixed interval of time during which the circuitry stores the input bit. To signal a logical 0, first the positive electrode is grounded and then the ground electrode is raised to V_{dd}. Again, this condition is maintained for a sufficient time to allow the circuitry to latch the input bit before returning to the power phase (P1 high and P0 grounded).

The bridge rectifier in the power circuitry charges a capacitor (and the rest of the power-up circuitry) regardless of which electrode is positive. The circuitry will function properly as long as the data and clock steps (step 3 above) are short compared to the power-up time constant (i.e., the time required to charge the power circuitry). This circuitry is useful because it works without regard to which electrode is positive and which is grounded.

If the assembled structures are to be deposited from suspension, they should be encased in a way similar to that illustrated in Figure 6.8. This structure connects opposing sides of a cube (or rectangular solid) to the P0 and P1 wiring inside the structure. Since opposite sides of the cube are connected

The Distributed Array Multiprocessor

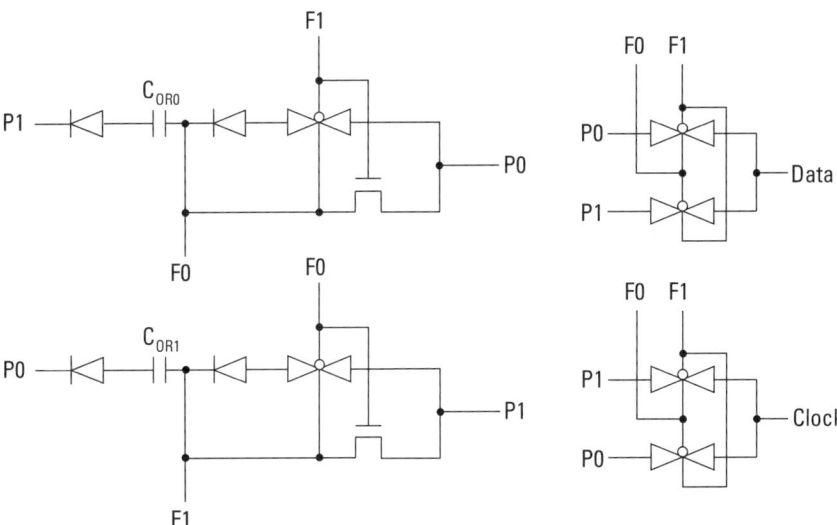

Figure 6.7 The power-up circuitry used to orient a structure after it has been sandwiched between two electrodes. The circuitry tells the structure which electrode was powered up first and therefore which electrode will serve as a data signal. The other electrode is the implied ground and clock signal.

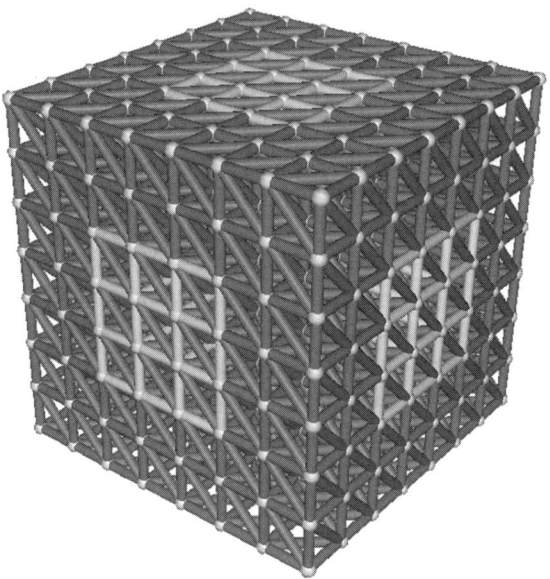

Figure 6.8 This structure connects opposing sides of a cube (or rectangular solid) to the P0 and P1 wiring inside the structure so that the structure can land with any side down and still receive electrical power and be able to communicate.

either to P0 or P1, the structure can land on a metallic surface (the bottom electrode) with any of its sides. A layer of insulating material (e.g., a polymer) could be deposited onto the surface and etched back to expose the top side of each cube. Another layer of metal could then be deposited on top and used as the positive electrode (see Figure 6.6). This method can be used regardless of how the cube lands as long there is Ohmic contact to the bottom and top electrodes.

6.4.3 Output Methods

Regardless of which interconnection method is used, there must be a way for each processing element to communicate with the outside world. It may be possible using future versions of self-assembling technology that have higher yields to make a single monolithic device with many connections between elements. In the near term where a single large monolithic device is not feasible and connections are restricted, a modular approach must be adopted.

With no connection other than a common power supply, each processing element is isolated. Aside from receiving commands there must also be a method for communicating calculation results. One potential solution to this problem is to use a switching-noise detector in the power supply. If the detector is tuned to a unique electrical oscillation made by the processing elements (called the ringer circuit) the system can output at least a single bit of information. The nature of this method makes the resultant bit take on the superposition of all bits being transmitted, which complicates the communication; only one processing element should communicate at a time.

As mentioned, one drawback of this method is the large capacitive load between the power electrodes and the frequency the voltages must change. That is, a single PE will have trouble overcoming the large capacitance of the power electrodes to produce a signal large enough to detect. This is a design trade-off between the number of PEs per unit area of electrode or the size of a PE and density. One solution to the capacitive loading problem is to design the PE to use a resonant communication method that relies on amplification at the node controller. A simpler way around this problem is to use conventional I/O methods to clock data into the processing element.

Figure 6.9 illustrates a modified footprint that has ports for clock and data signals that are shared in parallel by several processing elements. The ports are signal lines that protrude upward from the silicon substrate and through the passivation layer (e.g., glass). This is more complex than a simple metallic substrate because it needs multiple routing layers and control logic.

A processing element can directly sense the data line (or port) by wiring it to a multiplexer input. If the photolithographic process can place multiple

The Distributed Array Multiprocessor 89

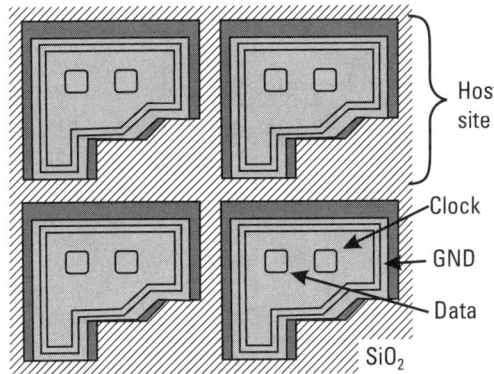

Figure 6.9 Modified footprint that provides a low capacitance clock and data port. The two squares inside the footprint are the data and clock electrodes; the ground electrode lines the rim of the footprint.

viaducts to the substrate the data channel can be expanded to multi-bit and/or full duplex. The external clock line can be driven by external amplifiers and connected directly to the internal PE clock line.

For output, external circuitry pre-charges the shared data line and any PE can pull the line to ground with pull-down logic. The low drive current of the nanoscale devices means that pulling the data line to ground could take a long time if the data line capacitance is large. The capacitance of the data line can be reduced by using an H-tree arrangement, illustrated in Figure 6.10, with buffers placed at vertices. The nodes on the H-tree sit below the modified footprints from Figure 6.9 and contain pre-charge control circuitry. The buffers pass the result of the pull-down to higher levels in the H-tree and ultimately to the top of the H-tree, or the processor node.

Since the data line may take a long time to pull to ground with the low drive current of the devices, it is necessary to keep the data line capacitance below a critical value. Using the decay constant $\tau = R \cdot C$ and assuming the line can be sensed after $3 \cdot \tau$ (or after > 90% of the charge has dissipated), the maximum data line capacitance C can be calculated. A drive current of 1.6 μA (i.e., a feasible drive current for a nanowire FETs) and an operating voltage of 1V implies an on-state resistance of 625 kΩ, which is the R in the decay constant. Solving for the capacitance, $C = \tau / R$ where τ is defined by the operating frequency. Operating at 400 MHz the total decay time $3 \cdot \tau = 2.5$ ns, or $\tau = 0.83$ ns. With the on-state resistance this capacitance establishes a maximum data line capacitance of 1.3 fF which is practical for standard CMOS technology. (A 0.25 × 30-μm plate capacitor with a 0.25-μm separation and $\kappa = 5 \cdot \varepsilon_0$ also has a 1.3 fF capacitance).

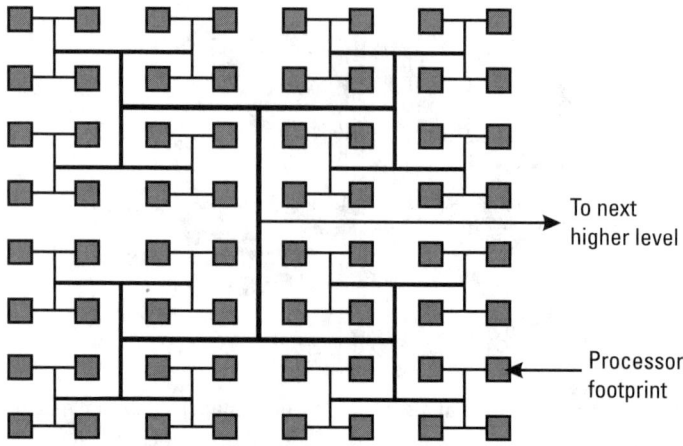

Figure 6.10 A portion of the processor node H-tree. This circuitry lies beneath the processing elements (and the passivation layer) in the silicon substrate. Each processing element footprint is ~4.5 μm on a side.

6.5 System Operation

The large number of processing elements in the DAMP ($\sim 10^{12}$) leads to a very high peak performance on 16-bit operands (e.g., local 16-bit adds) as illustrated in Figure 6.11. However, peak performance is only a simple measure of usefulness.

The size of the DAMP enables it to solve vast global optimization problems. Many science and engineering problems can be posed as global optimization problems that seek to find the largest or smallest value for an objective function over a domain. The challenge in solving these problems comes from the large number of variables and local minima that deceive search algorithms. Consider the hypothetical objective function shown in Figure 6.12. This function has many local minima, and the global minimum, indicated by the black arrow, has a very narrow aperture for the search algorithm to find.

Stochastic global optimization is a method of sampling an objective function at random points in the problem space and comparing the results at each point. Since the time required to exhaustively search the problem space is too large, the best local solution is chosen from local searches starting at a random set of locations. A new set of random points is selected that concentrates the search around the best-found solution(s). That is, the search continues but focuses on a few of the last round's best answers. Typical calculations for each sample include the objective function and numerical derivatives (gradients) at the point. If the objective function has a well-behaved (i.e., contin-

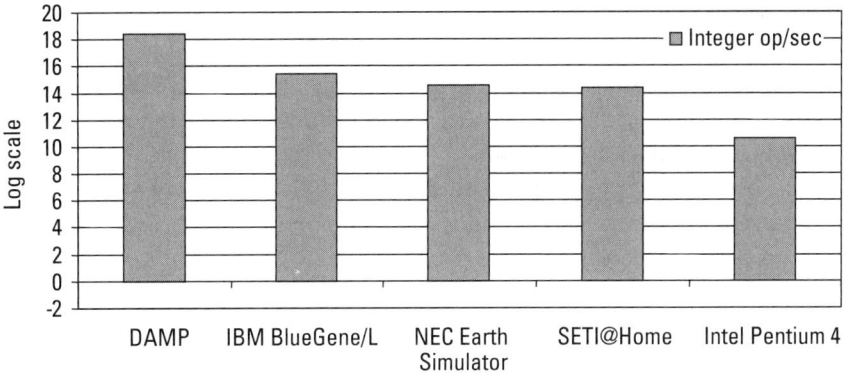

Figure 6.11 Peak performance of several machines on 16-bit operands.

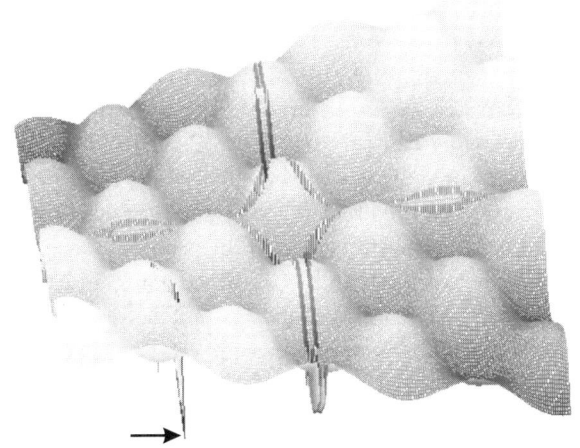

Figure 6.12 A constrained objective function with many local minima. The black arrow indicates the minimum for this region.

uous) and computable gradient, this can be used as a local indicator of how to choose the next best solution since the objective changes along the gradient. However, the gradient descent approach is very sensitive to numerical instability because of how derivatives amplify high frequency changes in the objective function. The gradient descent approach is also very susceptible to entrapment by local minima.

Parallel pattern search (PPS) is a technique used to optimize difficult objective functions [12]. This method uses a search along each dimension of the problem space to find the globally optimal solution. An iteration of the search begins from the optimal point found in the last round of evaluations. This technique has provable convergence to the optimum as long as certain rules are followed for adjusting the step size along each dimension and for how objective values are compared.

This approach to global optimization has been applied to continuous and mixed-variable problems and the typical continuous-variable optimization problem is formulated as follows:

```
minimize y = F(x)
where    x = (x₁, x₂, x₃, ..., x_N) ∈ X, and
         y ∈ Y.
```

The formulation above can be restated as a minimization of a function F that is subject to input and output constraints. That is, x_1 through x_N must belong to an allowable set of inputs X, and the output y must belong to an allowable set of outputs Y. Generally, N is less than 50 and several thousand iterations are required to converge to the best-known solution [10–12, 18].

The DAMP can be used to solve continuous variable minimization problems that are much larger in dimensionality than those solvable today. The pseudo code below can be used with 32-bit fixed-point variable optimization problems. The vector x_k is the best-known solution after each step. The problem space is spanned by a positive spanning set D, where d_i is a unit vector from D along the ith dimension of the problem space. A positive spanning set is a set of vectors that can be combined using non-negative scalars to form all possible vectors in a constrained space. The functions $C_x(x_i)$ and $C_y(y)$ are used to verify that the input and output vectors, respectively, satisfy the problem constraints.

The program listed below is run at each processing element. Since there are 2^{28} processing elements per node, the random number generated at each processing element has only 28 bits of significance. This means that to cover a 32-bit random number space each processing element must run the program 16 times with a new 4-bit low order value each time. The value, Δk, is simply incremented between loops. Each processing element takes its Δk and uses it to compute a new input vector. The particular dimension that the processing element searches along (d_i) is defined by the processor node. The new input vector is checked against the input constraints and if they are satisfied the function (F) is evaluated. The output from the objective function is checked against the output constraints and if they are satisfied the processing element participates in a minimization query, or MIN-QUERY. This query is conducted by the processor node and it searches, bit by bit, for the best objective function value found by any of its constituent processing elements.

Using PPS, the DAMP can solve continuous variable minimization problems that are much larger in dimensionality than those solvable today. The pseudo-code below demonstrates how PPS can be used to solve 32-bit (input space) optimization problems on the DAMP.

```
NODE CONTROLLER: For each processor node j (PNj),
NODE CONTROLLER: For each processing element at node PNj
NODE CONTROLLER: choose a fixed kj = random 28-bit integer
1: Δkj = kj << 4;

NODE CONTROLLER: For 16 iterations,

2: evaluate and verify Cx(xk + dj · Δkj)
3: evaluate y = F(xk + dj · Δkj);
4: evaluate and verify Cy(y)
5: participate in MIN-QUERY(y, xk + dj · Δkj)
6: Δkj = Δkj + 1;
```

The first *for* statement must be executed sequentially on the DAMP. Once the program has been run, the X_k solution vector that best minimizes the objective function is chosen for the next round.

The second *for* statement in the program can be distributed in parallel to the 4,096 processor nodes available on the DAMP. The third *for* statement can be run in parallel on all the processors within a node because they each use the same d_i vector in calculating a new X_k vector. This means that the DAMP can, in parallel, optimize a 4,096-dimension (M = 4,096) 32-bit fixed-point problem per round of the program above.

The following discussion provides an analysis of the program's execution time.

```
1: Δk = k << 4;
```

Step 1 first requires the random integer k to be loaded. The LOAD instruction can be used for this. The multiplication can be implemented by either performing four logical left-shifts, or a multiword circular right-shift by 28 bits followed by a logical AND with 0xFFF0. This step will take no more than 2,000 cycles.

```
2: evaluate and verify Cx(xk + dj · Δk)
3: evaluate yi = fi(xk + dj · Δk);
4: evaluate and verify Cy(y)
```

The constraint functions and objective function need to be preprocessed before being executed on each processor within a node to fit within the memory limitations. Since each processor node is responsible for a single search dimension, it is possible to precompute the value of all terms involving variables other than X_j. The other variables can be combined to form a function of the single variable X_j since all the other variables will remain constant at the jth processor node. If the precomputation of the single-variable form of the constraint and objective functions is considered to be part of the problem description, then only the complexity of the resulting functions is needed to estimate the execution time of step 2.

If the constraint and objective functions can be decomposed into a sum of terms that does not exceed the number of variables in the problem, M, then this is an estimate for the evaluation complexity. Each term requires a precomputed multiplier that represents the contribution of the other variables and will only need to have the contribution of X_j calculated. Each processor must perform this calculation since X_j is incremented by Δk, which is a processor-specific value. The calculation $X_j = X_j + \Delta k$ will not require more than 300 cycles. If each term requires 64 32-bit additions (316 cycles each) and multiplications (10,112 cycles each), then the entire function (4,096 terms) will require 2.73×10^9 cycles, or 6.8 seconds if the DAMP is at 400 MHz. That is, each constraint function, C_x and C_y, and the objective function, $F(x)$, has an estimated execution time of 6.8 seconds. Therefore, steps 2 to 4 will require $3 \times 6.8 \approx 20$ seconds on the DAMP.

```
5: participate in MIN-QUERY(y, x_k + d_j · Δk)
```

MIN-QUERY (a, b) is a routine that first queries the processors within a processor node for the minimum value, a, and then collects the argument, b, that generated the minimum value. The DAMP has enough register memory to store both the y value and the $x_k + d_j \langle \Delta k$ argument and participate in a 32-bit MIN-QUERY. The value being minimized, y, is first loaded into the accumulator and R0. This pair is then used in two 32-bit value binary searches (see Section 6.6) that require a total of ~44,000 cycles.

```
6: Δk = Δk + 1;
```

Step 6 is a simple 32-bit increment of the current Δk, which will not require more than 340 cycles.

```
7: repeat 16 times from step 2.
```

The total program from step 2 up to step 7 requires $\sim 2.7 \times 10^9$ cycles to execute, or 4.3×10^{10} cycles after looping 16 times. Therefore, the DAMP can completely sample an objective function with 4,096 dimensions at 32-bit resolution and return the best solution in 110 seconds.

In addition to the implementation details of each instruction, a practical application of this method to the thermal intercept problem and performance results based on ring-gated field effect transistors (RG-FETs) can be found in [2].

6.5.1 A DAMP Instruction Set

This section provides a detailed accounting of the cycles, execution time, total energy consumed, and maximum sustainable clock rate (for the DAMP with 10^{12} processor and a 3.5×10^6 W power budget) for several example instructions that can be found in Table 6.2.

The data for Table 6.2 were derived from simulation runs for each instruction using nanoscale device simulation results in a custom behavioral simulator. Figure 6.13 illustrates a typical output plot from the simulator. This plot shows how power is consumed over time and the total energy consumption for the execution of a single instruction.

The list of instructions in Table 6.2 is not comprehensive. Other instructions are possible by using SETSREG and SETCREG. SETCREG is used to set the target control register that SETSREG will eventually be used to fill. The CYCLE instruction is then used to step each processing element through the required number of cycles. The particular constants required to implement an instruction are defined by the microarchitecture and are described elsewhere [2].

Each operation, if it is not a variable bit-length operator, operates on the 16-bit accumulator with (or without) a single 16-bit register input. Instructions ending with the letter I indicate that an immediate value is used. The suffix letters GE, NGE, LT, and NLT signify greater-than, not greater-than, less-than, and not less-than, respectively. The SET* instruction must have a suffix of one of the inequality operators. The WAIT instruction must have a suffix to indicate which status bit to use from the following list: B, NB, C, NC, D, ND, S, NS, GE, NGE, LT, or NLT. The suffix indicates which status bit is consulted before setting or not setting the W status bit.

The SETGE, SETLT, SETNGE, and SETNLT instructions are helpful in restoring the wait status bit just before a RESUME instruction. The GE, LT, NGE, and NLT suffixes mean greater-than-or-equal, less-than, not greater-than-or-equal, and not less-than, respectively.

Table 6.2
Basic Instructions and Cycle Counts, Execution Time at 400 MHz, Energy Consumed, and Estimated Maximum Sustainable Clock Rate for 10^{12} Processors Operating with a Power Budget of 3.5-MW

Instruction	Cycles	Execution Time at 400 MHz (μs)	Total Energy Consumption (J)	Max. Clock Rate (MHz)
ADD	159	0.3975	1.2e−12	464
ADDC	131	0.3275	1e−12	459
ADDI	170	0.425	1.2e−12	496
ANDI	136	0.34	9.75e−13	488
ASR	116 + N	0.29 + N * 0.0025	8e−13 + N * 7.5e−15	502 (16 bits)
CLEARB	77	0.1925	6e−13	449
CLEARC | D	55	0.1375	3.75e−13	513
CMP	131	0.3275	9.5e−13	483
CMPI	307	0.7675	2e−12	537
CMPI8	211	0.5275	1.4e−12	528
COPY	92	0.23	7.5e−13	429
COPYH	215	0.5375	1.6e−12	470
COPYL	154	0.385	1.2e−12	449
COST	120	0.3	9.5e−13	442
CSR(N)	88 + N	0.22 + N * 0.0025	6e−13 + N * 9.375e−15	485 (16 bits)
CYCLE(N)	N	N * 0.0025	~N * 1.125e−14	311 (16 bits)
DEC	159	0.3975	1.1e−12	506
GRAB(N)	157	0.3925	1.1e−12	500
INC	159	0.3975	1.1e−12	506
LOAD	280	0.7	1.9e−12	516
LSR	88 + N	0.22 + N * 0.0025	6e−13 + N * 9.375e−15	485 (16 bits)
LSRC	66 + N	0.165 + N * 0.0025	4.5e−13 + N * 9.375e−15	478 (16 bits)
MCOPY	103	0.2575	8.5e−13	424
NOT	171	0.4275	1.2e−12	499
ORI	202	0.505	1.4e−12	505
RANDOM	55	0.1375	4e−13	481
RESUME	6	0.015	4.5e−14	467
RINGOFF	11	0.0275	7.2e−14	535
RINGON	11	0.0275	7.5e−14	513
SET*	55	0.1375	3.75e−13	513
SETB	77	0.1925	5e−13	539
SETCREG	6	0.015	4.5e−14	467

The Distributed Array Multiprocessor   97

Table 6.2
Continued

Instruction	Cycles	Execution Time at 400 MHz (μs)	Total Energy Consumption (J)	Max. Clock Rate (MHz)
SETC	55	0.1375	3.75e−13	513
SETSREG	5	0.013	4.2e−14	417
STORE	109	0.2725	9e−13	424
STOREH	171	0.4275	1.3e−12	460
STOREL	109	0.2725	9e−13	424
WAIT*	12	0.03	8e−14	525
XOR	148	0.37	9.25e−13	560

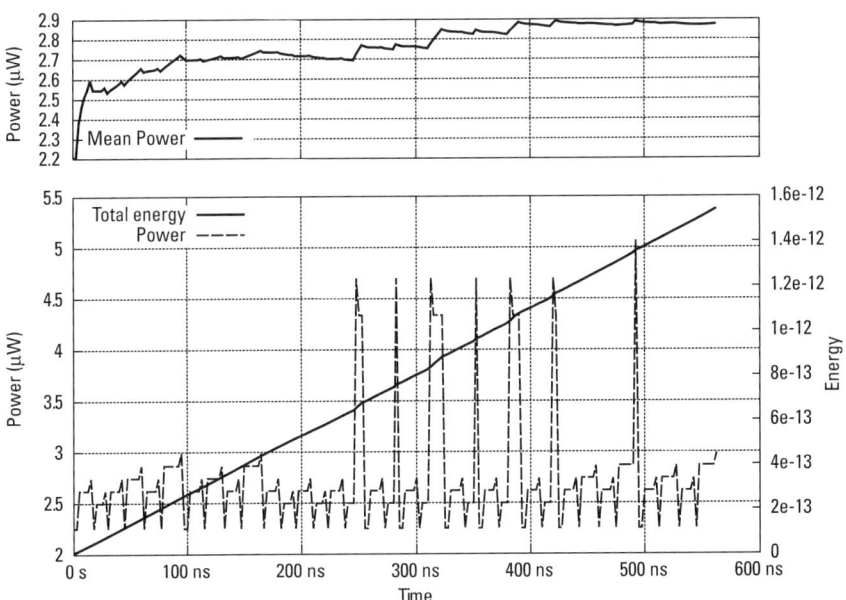

Figure 6.13 Output plot from the behavioral simulator for the ADDI instruction.

As with other SIMD machines, a simple IF-THEN-ELSE clause can be implemented as follows:

```
CMPI(constant)    // if(ACC >= constant) {
WAITNGE           // All PEs failing the condition WAIT...
...THEN-clause instructions...
SETGE             // restore the status bit
```

```
RESUME              // All PEs resume execution...
WAITGE              // All PEs that had previously executed
                    // the THEN-clause WAIT...
                    // } else {
...ELSE-clause instructions...
RESUME              // } // All PEs RESUME execution...
```

Nested IF-THEN-ELSE statements can be implemented in a similar manner as long as the corresponding SET* and WAIT* instructions are executed before and after each clause.

6.6 Performance

The average number of cycles in a DAMP integer operation was obtained by averaging the number of cycles used by the ADD, ADDI, ADDC, NOT, INC, and DEC. The average obtained using this instruction distribution is 158 cycles, or 0.395 μs per integer operation at a clock rate of 400 MHz. This instruction distribution consumes an average of 1.1×10^{-12} J per instruction. Therefore, the maximum clock rate for integer operations under a 3.5-MW power budget is 502 MHz. This is well above the conservative 400-MHz clock rate used here.

The results of intermediate computations within the DAMP may need to be communicated to the node controller. A simple binary search can be used to find the identity (i.e., the random constant) of a processing element that has calculated a value of interest (e.g., an extrema). The following code implements a binary search for the identity of the processing element with the largest value in its 16-bit accumulator:

```
NODE CONTROLLER: i = 32768  // initial pivot for search
NODE CONTROLLER: n = 2      // initial level in search
NODE CONTROLLER: do {

1: CMPI(i)   // Compare the ACC value to i (the current pivot)
2: WAITLT    // Any PE less than the pivot should wait
3: RINGON    // Used to detect the presence of PEs
             // with ACC values greater than i
4: RINGOFF

NODE CONTROLLER: If the RINGER signal was detected during
                 step 3
NODE CONTROLLER: i = i + (32768 / n), n = n * 2
```

NODE CONTROLLER: Else (no RINGER signal detected during
                step 3)

5: RESUME   // Need to wake all the PEs (i is too big)

NODE CONTROLLER: i = i - (32768 / n), n = n * 2
NODE CONTROLLER: } while n < 32768 && 0 < i < 65536

// At this point, i-1 is the largest ACC value in the DAMP
6: RESUME
7: CMPI(i - 1)
8: WAITLT // Only the largest ACC holder(s) remain active

// Time to determine the constant... (destroys ACC, R0, R1
   value)
9: CLEARB   // The B-bit will be used later to single out
              PEs
10: LOAD   // Load the constant
11: CSR(1) // Copy the first bit into the S-bit
12: WAITNS // If the bit is a 0, wait.
13: RINGON
14: RINGOFF

NODE CONTROLLER: If RINGER signal detected during step 13
NODE CONTROLLER: Record possible 1 in current bit position

15: RESUME
16: WAITS   // If the bit is a 1, wait.
17: RINGON
18: RINGOFF

NODE CONTROLLER: If RINGER signal detected during step 17
NODE CONTROLLER: Record possible 0 in current bit position

// Multiple unique processors with the same ACC value
// may respond and make both a 0 and a 1 the likely
// current bit. That is, the RINGER signal will be
// detected both times. In this case, each value must
// be pushed onto a stack and one chosen to proceed
// with the binary queries (disabling any PE with the
// other bit value.) Once a word is completed,
// the process must be repeated for the other bit value,
// recursively, unless only one solution is required.
// The procedure here assumes a single solution is needed.

```
19: RESUME


NODE CONTROLLER: If the 0 is the current bit value
NODE CONTROLLER: (i.e., step 17's signal > step 13's)

20: WAITNS   // The 0 bit PEs wait
21: SETB     // The B-bit represents "do not participate"

22: RESUME

NODE CONTROLLER: If the 1 is the current bit value
NODE CONTROLLER: (i.e., step 13's signal > step 17's)

23: WAITS    // The 1 bit PEs wait
24: SETB     // The "do not participate" flag

25: RESUME
26: WAITB    // Any PE with the B-bit set will wait

NODE CONTROLLER: Append the current bit to the accumulated
                 constant
NODE CONTROLLER: Repeat from step 11 for the remaining 15
                 bits
NODE CONTROLLER: Repeat this entire procedure twice more
NODE CONTROLLER: for R0 and R1
```

Steps 1 through 5 search for the largest ACC value and require (in the worst-case) 260 cycles per loop. The loop is executed 16 times for a total of 4,160 cycles. Steps 11 through 26 require 355 cycles per loop. Since this loop must be executed once for each random bit (48 times in total) the total number of cycles for just this loop is 17,040. The entire procedure requires 21,200 cycles, or 42.4 μs using a 400-MHz DAMP.

6.6.1 Simple Comparisons

The number of integer operations per second, memory size, power consumption, and volume of each machine can be used to draw a simple comparison among systems. Such a comparison gives a qualitative feel for how the machine may perform on simple tasks. Table 6.3 lists several simple metrics for comparison.

The data for Table 6.4 was taken, and in some cases extrapolated, from several architectural overviews and surveys [13–15]. The 16-bit integer

The Distributed Array Multiprocessor 101

Table 6.3
Comparison of Several Machines with Respect to Integer Operation Rate,
Memory Size, Power Consumption, Volume, and Energy per Operation

Machine	16-Bit Integer op./sec	Norm. Integer op./sec	Memory Size (Bytes)	Power Budget (W)	Volume (m^3)	Energy / op. (J/op)
DAMP	2.53×10^{18}	1.00	1.0×10^{13}	3.5×10^6	12	1.3×10^{-12}
IBM BlueGene /L	2.88×10^{15}	1×10^{-3}	7.0×10^{13}	$\sim 3 \times 10^6$	533	1.04×10^{-9}
NEC Earth Simulator	3.28×10^{14}	1×10^{-4}	1.0×10^{12}	12.8×10^6	13,000	3.9×10^{-8}
SETI@home	2.4×10^{14}	9×10^{-5}	$> 1.4 \times 10^{12}$	508.3×10^6	312,800	2.1×10^{-6}
HP ASCI Q	6.2×10^{13}	2×10^{-5}	1.3×10^{13}	3×10^6	11,300	4.8×10^{-8}
IBM ASCI White	5.6×10^{13}	2×10^{-5}	2.6×10^{13}	1.2×10^6	6,600	2.1×10^{-8}
Thinking Machines CM-200	4.5×10^{10}	1×10^{-8}	8.59×10^9	28×10^3	4.16	6.2×10^{-7}
Intel Pentium 4	3.44×10^{10}	1×10^{-8}	$< 2 \times 10^9$	~ 150	0.08	4.36×10^{-9}
MasPar MP-1	1×10^{10}	3×10^{-9}	2.68×10^8	3.7×10^3	0.95	3.7×10^{-7}

Table 6.4
Time to Beat the Largest SpaceComputation on Record

Machine	Time to Beat the Largest Computation on Record	Approximate Normalized Time
DAMP	5 hours, 9 minutes	1.0
IBM BlueGene /L	207 days	800
NEC Earth Simulator SETI	4 years, 7 months, 23 days	7,000
HP ASCI Q	24 years, 7 months, 4 days	40,000
IBM ASCI White	27 years, 2 months, 23 days	~40,000
Thinking Machines CM-200	33,887 years	50,000,000
Intel Pentium 4	44,328 years	70,000,000
MasPar MP-1	152,496 years	200,000,000

performance for the Intel Pentium 4 was calculated using its sustainable memory to processing element transfer rate of 4.3 Gbps.

The Association for Computing Machinery (ACM) has the SETI@home project on record as the owner of the largest computation ever (being) performed with approximately > 10^{21} 64-bit integer operations. That number translated into the equivalent number of 16-bit operations, listed for each machine in Table 6.4, puts the scale of self-assembled computing into perspective.

6.6.2 Blind Decryption of the Data Encryption Standard

The data encryption standard (DES) has long been used as a secure method of encrypting sensitive information. Recently, the growing number of computing machines that can complete brute force attacks on secured data has put the strength of this form of encryption into question. For this reason alternative encryption standards are being investigated.

The purpose of this comparison is to show how the DAMP would perform on a particular problem of interest to a wide community. The DES has been heavily investigated since the late 1970s and its implementation is well known [16]. Fast bit-serial implementations have also been developed in an attempt to improve the throughput of DES encryption and decryption modules and to test the security strength of the standard [17].

The particular bit-serial DES algorithm implemented in [17] uses approximately 17,000 bit-wise operations per DES decryption. This count includes the comparison operation that tests whether or not the decryption was successful. The usual completion criterion for the algorithm is to successfully decrypt a single known ciphertext to the proper plaintext. In this case, the algorithm is being employed to perform a blind search through the entire DES key space to test the algorithm's security rather than perform a complete data stream decryption.

The DAMP can be used to perform this search by assigning each processing element a key from the 2^{56} bit key space. On average this input space will require 2^{55} decryption attempts. Since the DAMP has only 40 assembly-time bits, the remaining 15 bits must be obtained at run-time. The most common instructions in the algorithm are the XOR and CSR instructions. On average, the logic instructions in the DAMP require 10 cycles per bit-wise operation. Using this estimate, a single DES decryption will require 170,000 cycles on the DAMP.

The simulations and performance estimates use a clock rate of 400 MHz to determine the total running time. The DES decryption rate is 2,352 decryptions per second per processing element. The aggregate performance of

Table 6.5
Comparison of Blind DES Decryption Times for Various Machines

Machine	Bit-wise op./second	Avg. DES time	Normalized time
DAMP	4×10^{19}	15 sec	1
IBM BlueGene /L	4.6×10^{16}	3 hrs, 41 min	870
NEC Earth Simulator	5.25×10^{15}	32 hrs, 24 min	7,600
SETI@home	3.84×10^{15}	1 day, 20 hrs, 18 min	10,400
HP ASCI Q	9.9×10^{14}	7 days, 3 hrs, 51 min	~41,000
IBM ASCI White	8.96×10^{14}	7 days, 21 hrs, 53 min	~45,000
PixelFlow (CipherFlow)*	1.04×10^{13}	1 year, 10 months, 10 days	~3,840,000
Thinking Machines CM-2000	7.2×10^{11}	26 years, 11 months, 21 days	~56,000,000
Intel Pentium 4	5.5×10^{11}	35 years, 3 months	~73,000,000
MasPar MP-1	1.6×10^{11}	121 years, 4 months	~254,000,000

* Data from Kedem, 1999.

the DAMP, with 2^{40} processing elements, is approximately 2.5×10^{15} DES decryptions per second. At this decryption rate the DAMP can cover the entire 55-bit input space in about 15 seconds, or the entire space in 30 seconds. Table 6.5 lists the estimated performance of several machines on the blind DES decryption problem.

6.7 Summary

The DAMP is one example of a computer architecture that can leverage the large manufacturing scale offered by self-assembly. The key design feature is simplicity in the per-processor logic and interfaces. While simulations of the DAMP clearly demonstrate an advantage over other systems, not all problems can be efficiently solved by this architecture. However, global optimization is found in many scientific and engineering applications and this is a problem the DAMP can solve efficiently. The DAMP is a demonstration that as self-assembly matures there is at least one feasible and useful system design along the way.

References

[1] Dwyer, C., et al., "The Design of DNA Self-Assembled Computing Circuitry," *IEEE Transactions on Very Large Scale Integration (VLSI) Systems*, Vol. 12, No. 11, 2004, pp. 1214–1220.

[2] Dwyer, C., *Self-Assembled Computer Architecture: Design and Fabrication Theory*, Ph.D. Dissertation, Chapel Hill: University of North Carolina, 2003.

[3] Adleman, L., "Molecular Computation of Solutions to Combinatorial Problems," *Science*, Vol. 266, 1994, pp. 1021–1024.

[4] Braich, R. S., et al., "Solution of a 20-Variable 3-SAT Problem on a DNA Computer," *Science*, Vol. 296, 2003, pp. 499–502.

[5] Clark, T. D., et al., "Self-Assembly of 10-um-Sized Objects into Ordered Three-Dimensional Arrays," *Journal of the American Chemical Society*, Vol. 123, No. 31, 2001, pp. 7677–7682.

[6] Clark, T. D., et al., "Template-Directed Self-Assembly of 10-um-Sized Hexagonal Plates," *Journal of the American Chemical Society*, Vol. 124, No. 19, 2002, pp. 5419–5426.

[7] Soto, C. M., A. Srinivasan, and B. R. Ratna, "Controlled Assembly of Mesoscale Structures Using DNA as Molecular Bridges," *Journal of the American Chemical Society*, Vol. 10, 2001, pp. 17–24.

[8] Schmid, A. W., et al., "Low-Surface-Energy Photoresist as a Medium for Optical Replication," *LLE Review*, Vol. 66, 1996, pp. 82–85.

[9] Martin, B. R., S. K. St. Angelo, and T. E. Mallouk, "Interactions Between Suspended Nanowires and Patterned Surfaces," *Advanced Functional Materials*, Vol. 12, No. 11–12, 2002, pp. 759–765.

[10] Zitzler, E., K. Deb, and L. Thiele, "Comparison of Multiobjective Evolutionary Algorithms: Empirical Results," *Evolutionary Computation*, Vol. 8, No. 2, 2000, pp. 173–195.

[11] Fieldsend, J., and S. Singh, "A Multi-Objective Algorithm based upon Particle Swarm Optimization, and Efficient Data Structure and Turbulence," *Proc. 2002 U. K. Workshop on Computational Intelligence*, 2002, pp. 37–44.

[12] Hough, P. D., T. G. Kolda, and V. J. Torczon, "Asynchronous Parallel Pattern Search for Nonlinear Optimization," *SIAM J. Sci. Comput.*, Vol. 23, No. 1, 2001, pp. 134–156.

[13] Dongarra, J., "Notes on the Earth Simulator," *Technical Report, University of Tennessee*, 2002.

[14] MacDonald, N. B., "An Overview of SIMD Parallel Systems—AMT DAP, Thinking Machines CM-200 & MasPar MP-1," *Technical Report, University of Edinburgh*, 1992.

[15] Warren, M. S., E. H. Weigle, and W. C. Feng, "High-Density Computing: A 240-Processor Beowulf in One Cubic Meter," *Proc. Proceedings of the IEEE Supercomputing Conference*, 2002, pp. 1–11.

[16] Feldmeier, D. C., "A High-Speed Software DES Implementation," *Technical Report, Computer Communications Research Group*, 1989.

[17] Biham, E., "A Fast New DES Implementation in Software," *Proc. FSE '97: Proceedings of the 4th International Workshop on Fast Software Encryption*, 1997, pp. 260–272.

[18] Audet, C., and J. E. Dennis, "Pattern Search Algorithms for Mixed Variable Programming," *SIAM Journal on Optimization*, Vol. 11, No. 3, 2000, pp. 573–594.

7

A Nanoscale Active Network Architecture

7.1 Introduction

There are numerous challenges to overcome in order to exploit the potential of DNA-based self-assembly. Although the oracle and DAMP architectures overcome the challenges of DNA self-assembly, and even exploit some of the new capabilities it presents, they are applicable to a very limited class of applications. This chapter presents a case study of initial work that seeks to support more conventional computing paradigms.

The rest of this chapter is organized as follows. The proposed architecture is presented in detail in Sections 7.2 to 7.9. An evaluation of the architecture using two illustrative examples is provided in Section 7.10, followed by a discussion of the architectural limitations in Section 7.11.

7.2 A General-Purpose Architecture for Self-Assembled Nano-Electronics

The nanoscale active network architecture (NANA) is an initial approach to address the issues raised in Chapter 4 and also provides a general purpose architecture that is compatible with the underlying fabrication technology. The architecture is like an active network [1] where execution packets, which contain instructions and operands, search through a loosely structured sea of processing and memory nodes for the functionality that they need at each step of execution. This architecture matches the technology characteristics since it (1) allows for differing node types with specialized functionality, (2) tolerates a random interconnection of nodes, and (3) tolerates node and interconnect fabrication defects.

7.3 System Model

The system model is a random interconnection of various node types, in which all nodes contain circuitry for communication and each node has some specialized circuitry (e.g., processing, memory, and so on). Groups of nodes are organized into cells. A node communicates with a neighboring node via a single link that is asynchronous and bidirectional (time-multiplexed on a single physical wire), illustrated in Figure 7.1. Each cell has a via that is its connection to the microscale, and one of the nodes connected to the via acts as the anchor node for the cell. Intercell communication occurs through a microscale interconnection network. The memory nodes in each cell comprise a portion of the global memory space. A fraction of the nodes are configured as memory ports to provide an interface between execution packets and memory storage. Figure 7.2 illustrates the system model. To impose structure on the interconnection network and the memory system, there is a configuration phase [2] that occurs before any execution. Reconfigurable architectures [3–6] have demonstrated that this approach is important to achieve high performance in the context of highly focused (i.e., aggressive) or highly defective technologies, including nanotechnology. A description of the purpose, beyond defect tolerance, and operation of the configuration is given in detail later in this section.

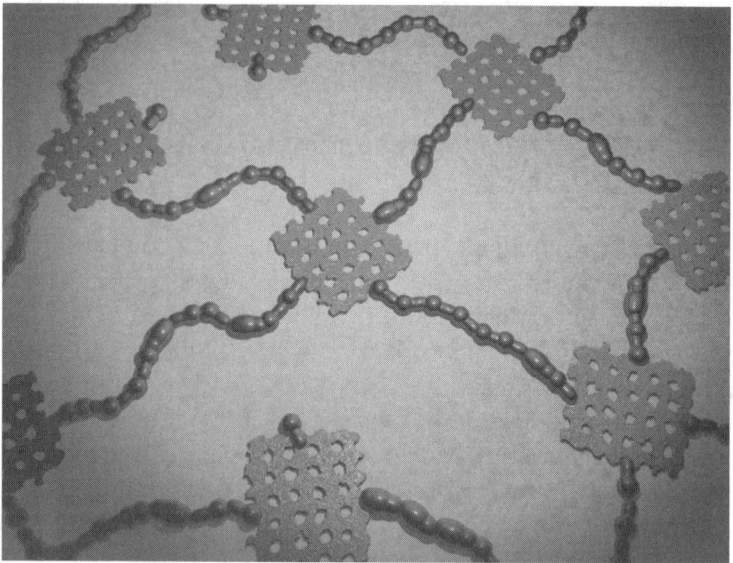

Figure 7.1 Schematic rendering of a self-assembled DNA interconnection network after metal deposition.

While node functionality is heterogeneous, all nodes have some common responsibilities. Each node generates its own local clock (this work assumes a clock frequency of 10 GHz, which is pessimistic given the data in Figure 7.3) and communicates asynchronously with its neighboring nodes

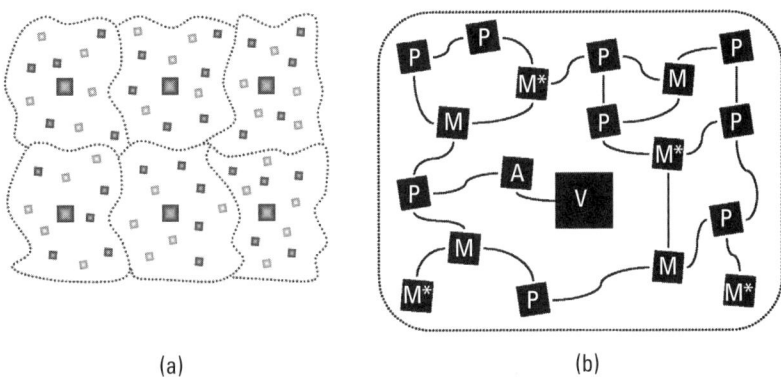

Figure 7.2 (a) System model. (b) Processing nodes (P), memory nodes (M), memory port nodes (M*), anchor node (A), and via (V). This schematic is not to scale (w.r.t. nodes per cell).

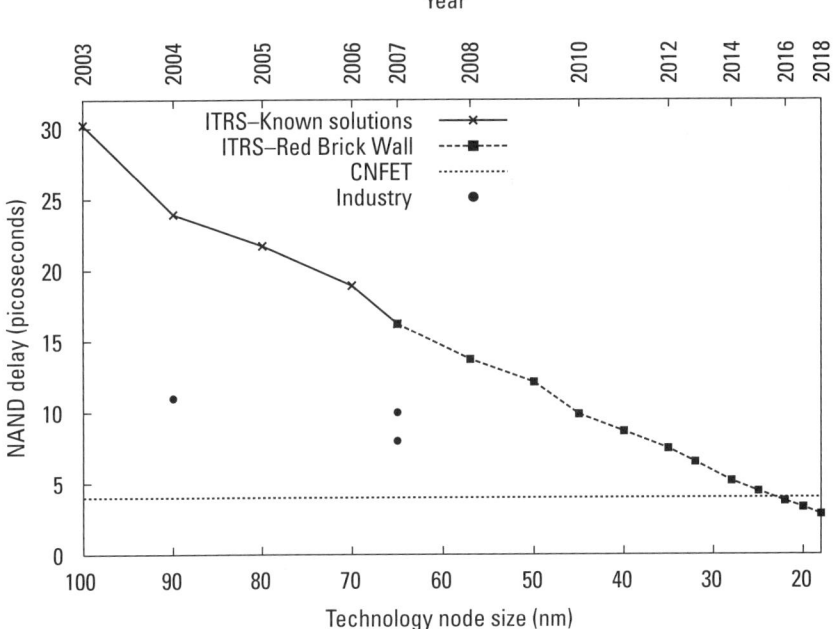

Figure 7.3 Nanoscale device performance.

using signaling techniques similar to push-style pipeline systems. High level communication between two devices over a single wire can be managed using simple two- and four-phase single wire techniques [7]. Each node must also contain routing functionality for determining the outgoing link for an incoming packet (or the result of an operation). This circuitry maintains node state (e.g., currently processing a packet) and handles link contention.

7.4 Execution Model

The execution model relies on an accumulator-based ISA. Conceptually, the accumulator is initialized and then a sequence of operations is performed on the corresponding series of operands. The accumulator-based ISA reduces the need for widespread a priori coordination and communication among many components (e.g., associative lookup in issue queues), since the only data dependence involves the accumulator and instructions are processed in order [8]. The architecture supports accumulator-based execution by forming an execution packet that contains the operations, the accumulator, and all operands in appropriate order. Instructions are executed in the order specified in the packet, as they are routed through the network and find the requisite functional units (or memory ports). Logically, each functional unit performs its specified operation, removes the operand and forwards the new accumulator and the remaining operands to the subsequent functional units. Each subsequent functional unit performs a similar sequence until all operations in the packet are completed. Memory operations generate memory packets that are handled by the memory ports, as discussed in Section 7.7. Packet sequencing is achieved using a process called chaining, discussed in Section 7.8.

The system and execution model enable significant parallelism by instantiating multiple execution packets within a cell and in multiple cells. While this parallelism is an important aspect of the architecture that fully exploits the capabilities of the underlying technology, this chapter focuses primarily on the operation of a single cell and sequentially instantiated execution packets.

To augment the defect tolerance of configuration and to protect against transient faults, one could add a signature vector to each packet and verify the integrity of a computation performed by the packet. The signature vector is operated on like the operands field of a regular execution packet with the exception that the initial signature is not consumed by the operation. The order of instructions will be reflected by a characteristic signature vector and can be used to determine if the nodes performing those operations were functioning properly during the signature calculation. This approach can be further augmented with redundant execution packets and a voting mechanism.

7.5 Instruction Set and Packet Formats

The format of an execution packet is: header, instructions, operands, tail. Specific bit patterns delineate field boundaries. The header is a fixed-length field that includes packet type and other metadata. The instructions field is a variable length list of opcodes in program order. The operands field is a variable length list of operand values. To accommodate the limited node size, this system uses a bit-serial implementation. The active network architecture and accumulator ISA are independent of this choice and provide an architecture that can scale with improvements in node capabilities (i.e., multibit operations). Figure 7.4 shows the execution packet format for the bit-serial implementation. The operands field is divided into bit-slices from least significant bit to most significant bit (from packet head to tail). Each bit slice starts with a bit from the accumulator and is followed by each bit (for the particular bit-slice) of the operands.

The instructions that this architecture supports must be bit-serial in nature and require little communication between bit slices. Many instructions are simple to implement with limited circuitry (e.g., ADD, SUB, OR, AND, XOR, NOR, NAND, compare, move) and require only small extensions to a bit-serial full-adder circuit. Each operation requires only a small amount of information (e.g., carry-out bit) to be communicated to subsequent bit slices. This simplifies the implementation details of the circuits so that they will fit within the node size limits of the technology. Although each instruction is bit-serial, the bit interleaving enables parallel execution of consecutive operations in a pipelined manner. Instructions supported by NANA can be divided into nine categories and are listed in Table 7.1.

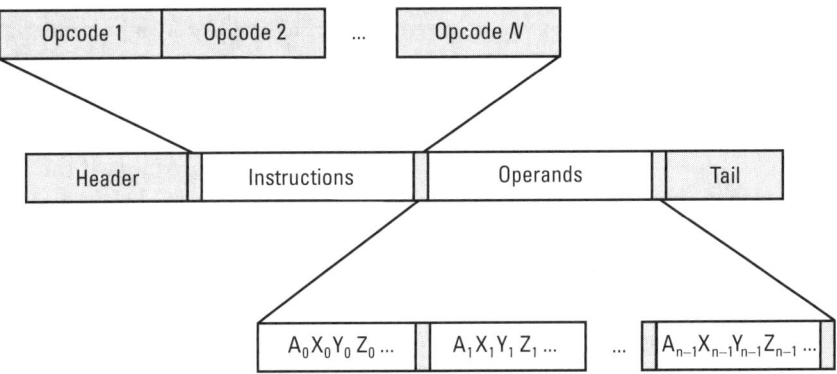

Figure 7.4 Packet format.

Table 7.1
NANA Instruction Set

Instruction Type	Instructions
Arithmetic	ADD, INC, SUB, DEC, SHL, SHR
Comparison	COMPEQ, COMPGT, COMPLT, SETEQ, SETGT, SETLT, SETZ
Operand stream control	LDCONST0, LDCONST1, CPACC, MOV, DELOP, OPFLUSH, SWAP
Logical	AND, NAND, NOR, NOT, OR, XOR, XNOR, NOP
Load	LD [Mem], LDI [Mem]
Store	ST [Mem], STI [Mem]
Conditional store	CST [Mem], CST_RST [Mem], CRST [Mem], CSTI [Mem], CSTI_RST [Mem], CRSTI [Mem]
Unconditional control transfer	JMP [Mem], CALL [Mem], JMPI [Mem], CALLI [Mem]
Conditional control transfer	CALLNZ [Mem], CALLZ [Mem], CALLNZI [Mem], CALLZI [Mem]

The serial nature of this architecture and the limited node complexity of the technology make certain operations difficult. Table 7.2 lists several instructions specially designed to help overcome these difficulties. For example, right shifts (moving bits from the tail toward the head) are difficult because they require bits to be forwarded ahead of other bits unless entire operands are stored at the functional node. Since supporting both operand storage and ALU-type functionality in a single node requires too much area for the limited node size, the architecture exploits the stack-like nature of the operand stream to support right shifts. When a right shift is executed, it also places the result at the end of the operand stream. Thus, to execute a right shift, the node buffers the field separator between bit slices and sends out the next observed data bit before reinserting the field separator into the packet bit stream.

The bit-slice packet encoding also complicates memory operations. For example, a load requires all of its address bits to generate a request. If the address is in the operand stream, then it is impossible for the load to interleave the resulting data in the same operand stream since all the low order bits are ahead in the packet flow before the entire address is obtained. Similar difficulties exist for stores. Therefore a packet cannot both calculate an address and use it in the same packet. To address these limitations, the system provides three specific types of memory addressing: immediate, constant address, and indirect address. Constant addressing requires the address to appear in the in-

Table 7.2
Definitions of a Selected Subset of Instructions

Instruction	Operation
MOV	Move accumulator to end of operand stream
SWAP	Swap first and second operand
SHR	Shift accumulator right by 1 bit, move accumulator to end of operand stream
DELOP/OPFLUSH	Remove one/all operands from operand stream
CPACC	Create copy of accumulator at end of operand stream
SET (EQ/GT/LT/Z)	Set flag bit in tail if condition satisfied, consume accumulator
COMP (EQ/GT/LT)	Set flag bit in tail if condition satisfied, consume first two operands
LDI [Mem]/STI [Mem]	Load/store indirect through constant address [Mem]
CST [Mem]/CSTI [Mem]	Conditional store direct (CST) or indirect (CSTI) to [Mem] (status bit in tail must be set)
CST_RST [Mem]	Conditional store to [Mem], reset status bit after performing store
JMP [Mem]/JMPI [Mem]	Fetch instructions into existing packet from direct (JMP) or indirect (JMPI) address [Mem]
CALL [Mem]/CALLI [Mem]	Create new packet using instructions from direct (CALL) or indirect (CALLI) address [Mem]
CALLNZ [Mem]/CALLNZI [Mem]	Fetch instructions into new packet if status bit is set (not zero) (direct/indirect)
CALLZ [Mem]/CALLZI [Mem]	Fetch instructions into new packet if status bit is not set (zero) (direct/indirect)

struction field of the packet. Indirect addressing supports indirection through a memory location that is specified as a constant in the instruction field of the packet. The architecture also includes special instructions (JMP & CALL) for instruction sequencing (discussed in Section 7.8). Conditional execution is supported through status bits (e.g., condition codes) in the packet tail. Currently the architecture includes conditional store and CALL instructions that must wait to execute until the packet tail arrives so that they can examine the appropriate status bit.

Programming NANA is similar to programming other accumulator-based ISAs [8–11]; however, care must be taken to account for system capabilities and constraints. For example, the "shift right" instruction (SHR) is constrained by node resources to shift the accumulator and move it to the end of the operand stream, while the "shift left" instruction (SHL) operates as expected (i.e., it shifts the accumulator left by 1 bit). Another constraint arises from the structure of the memory system: all loads must precede stores in a packet. Consider a simple code fragment ($x = x + {}^*(y + a)$) that computes a memory address ($y + a$) and then adds the contents of that location to another variable stored in memory. Due to the load-store ordering constraint, instructions must be divided into two packets. Table 7.3 shows the two packets needed to implement the code segment, and how their fragments are arranged in memory. The first packet, starting at address 0x10, performs an address calculation ($y + a$) and stores the result in a third location, z. The last instruction, at address 0x20, chains this packet to the next packet, which starts at address 0x40. The second packet performs the addition of x with the value stored at the memory location pointed to by z, and stores the result into x (i.e., $x = x + {}^*z$). This packet executes by first loading the value of x, then performing an indirect load on z (instruction at 0x44). Next, it executes the add and stores the result into x. This example illustrates some constraints that must be faced in programming NANA. As the underlying technology matures, a richer ISA with more complex instructions may become possible, including efficient variable bit shifts, bit-serial multiplication, and division. Until then, these more sophisticated operations must be composed in software using simpler primitives.

7.6 Interconnection Network: Finding Resources for Execution

The active network architecture must enable packets to find what they need without deadlocking or livelocking, despite high defect rates and traveling through a randomly interconnected sea of nodes. To avoid request/response deadlock (i.e., fetch deadlock), the minimum requirement is three logical networks: one for execution packets, one for memory request packets, and one for memory response packets. Each of these logical networks is irregular and must provide deadlock- and livelock-free routing. While three virtual channels [12] per unidirectional link would suffice to implement these three networks, this unnecessarily increases the amount of buffering required on a single node. The requirement can be reduced to two virtual channels per unidirectional link by creating distinct physical networks for execution and mem-

Table 7.3
Memory Layout for Two Packets that Compute $x = x + {}^*(y + a)$

Address	Instruction	NextPC	Address	Instruction	NextPC
0x10	LD y	0x14	0x40	LD x	0x44
0x14	LD a	0x18	0x44	LDI z	0x48
0x18	ADD	0x1A	0x48	ADD	0x4A
0x1A	ST z	0x1E	0x4A	ST x	0x0
0x1E	CALL(0x40)	0x0			

ory; the implementation of this is explained in Section 7.7. Wormhole routing is also used since it requires the least buffering on each node (1 bit per channel).

7.6.1 Imposing Structure with Gradients

Virtual networks avoid fetch deadlock, yet each network must still provide deadlock- and livelock-free routing. Given the irregular networks created during fabrication and due to defects, the system is first configured to create a spanning tree using the reverse path forwarding algorithm (RPF) [2, 13] and then employs a variant of up*/down* routing [14], a degenerate case of turn-model routing [15], and back pressure flow control. The challenge is implementing these techniques with limited node functionality.

To meet this challenge, each node is equipped with two forms of communication: (1) broadcast, and (2) routing along gradients [16, 17]. Packet headers include information on the type of communication to use. Broadcast requires minimal state per node and is used during configuration only. Gradients reduce per-node resources while still enabling deadlock- and livelock-free routing. The RPF algorithm creates a spanning tree with a specific via as the root and establishes a gradient with a specialized packet. Each node marks the link on which the gradient packet was received (i.e., points to its parent in the spanning tree) and broadcasts the packet to its other neighbors. A node will not broadcast gradient packets after having seen the first packet. This process can be generalized to any number of gradients if each node records an identifier for each gradient it detects. The broadcast algorithm terminates when all reachable nodes have received the gradient. There is no external action required to terminate the algorithm, and each node automatically stops forwarding broadcast packets when it has been configured.

There are five gradients: one for each planar direction (north, south, east, and west) and an additional gradient that establishes cell boundaries and the direction toward the via in each cell (called the cell gradient). The planar gradients are established by starting the broadcast at the north, south, east, and west edges (or corners) of the system, respectively. Figure 7.5 illustrates a gradient established from the upper left corner (north) in a 32 × 32 grid with a 30% defect rate. Defective nodes, not drawn in this figure, can cause islands of disconnected nodes such as the region near the via.

Configuring Cells. The process is initiated at each via in parallel by broadcasting a cell ownership packet that includes a cell identifier. The cell gradient broadcast stops when its wave front collides with the wave front from an adjacent via. Nodes detecting a conflict in cell identifiers stop the broadcast, creating a boundary between cells, and record that they are boundary nodes.

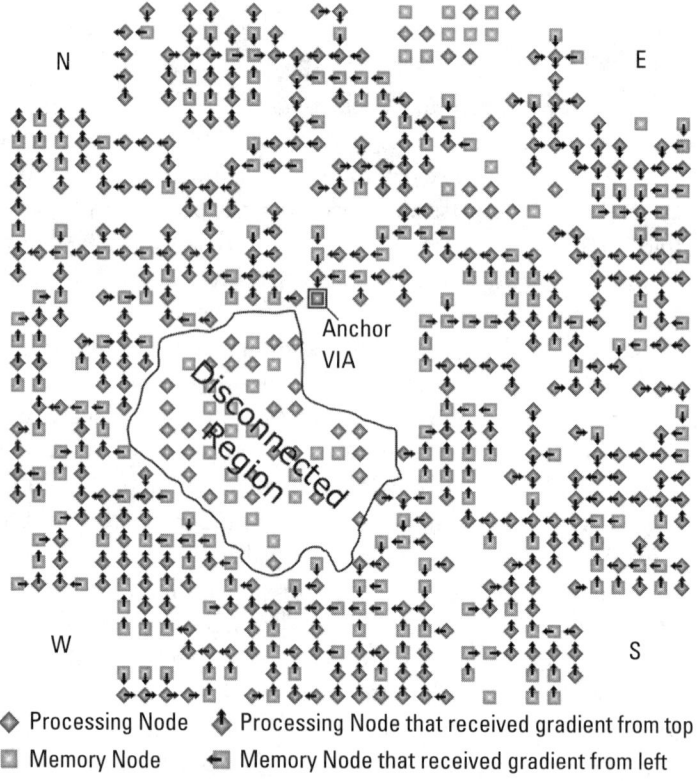

◆ Processing Node   ↥ Processing Node that received gradient from top
▯ Memory Node   ⬅ Memory Node that received gradient from left

Figure 7.5 A 32 × 32 grid of memory and processing nodes with one established gradient (north).

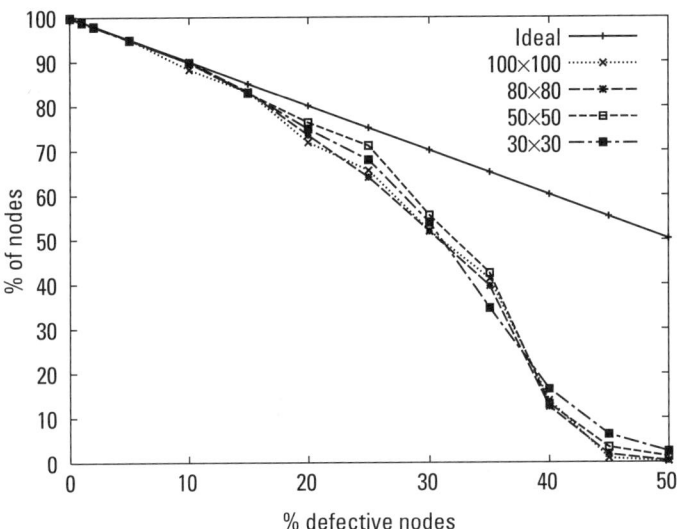

Figure 7.6 Percentage of nodes reachable by gradient broadcast for varying defect rates.

Tolerating Defects. Creating spanning trees using a broadcast flood maps out defective nodes and links, since no other node will have a gradient pointer to the defective node. If routing is restricted to follow a gradient, then packets will never be sent to a node that did not receive a gradient packet. Figure 7.6 shows the percentage of nodes that can be reached by a broadcast for increasing node defect rates. Each point on a curve is the average of ten simulation runs with different distributions of defects in the network. The different curves correspond to different network sizes. For defect rates up to 20%, the broadcast reaches most of the functional nodes. A majority of functional nodes is still reachable for defect rates up to 30%, beyond which there is a sharp drop in the number of nodes receiving the broadcast because increasingly large regions are isolated from vias. The analysis also shows that it is better to broadcast the planar gradients from an edge than from a corner of a rectangular or square network of nodes. In general, a via with more nodes surrounding it has a better chance of reaching a larger set of nodes.

The defect model assumes that the routing circuitry for a node is either fully functional or not operational at all.[1] The system can tolerate shorts in the node interconnect, called broadcast defects because they represent the

---

[1] The general Byzantine defect model, in which defective nodes can produce arbitrary behavior, has been considered in the Internet literature, but tolerating such defects requires a great deal of complexity at each node [18].

unintentional broadcast (to more than one link) of packet bits. Such defects are difficult to avoid during fabrication and require an arbitration scheme, similar to fixed back-off media access schemes in networks. The asynchronous link controllers in each node can be designed to assert a link-good signal after a random interval of time after power-up. The randomness can be introduced during the self-assembling process as in the oracles [19]. Every node monitors its links for the link-good signal and marks any link that has received more than one signal as defective. When the node's internal random interval has elapsed, if the link is not already marked defective it asserts its own link-good signal on all links. This arbitration scheme identifies both shorts and opens on links between nodes. The nodes connected to the via essentially share a single link (the via) which appears as a broadcast defect. The result of this arbitration scheme is for a single node to remain actively connected to the via, thus acting as the cell anchor.

Due to defects, some vias may not have a path to any of the four planar gradient destinations. This can be detected by monitoring the via at the microscale during the broadcast of each of the planar gradients. If the via fails to receive any of the gradient broadcast packets, it should be marked as defective and not participate in cell configuration.

7.6.2 Execution Packet Routing

The spanning tree structure imposed by gradients provides the framework for packet routing. Execution packets and memory packets never share physical links and thus cannot block each other. Up*/down* routing on the spanning trees prevents routing deadlock and livelock. However, execution packets must be able to find the necessary resources for execution, and memory packets must successfully find the appropriate memory location, which responds if necessary. Memory packet routing is discussed in Section 7.7. To avoid deadlocking execution packets, a packet simply follows a single gradient (up* on one spanning tree) on one virtual channel until it reaches a cell boundary, then the packet is reflected back into the cell on the opposite planar gradient but on the other virtual channel. Reflection only occurs if there are remaining instructions in the packet, otherwise a special packet is sent to the anchor node to indicate completion. Note that the header can run ahead of the operand stream allocating nodes for instructions (due to execution delay in a node). This approach can indefinitely bounce a packet between cell edges. The only constraint is that packet length be less than the total number of nodes in the round-trip traversal. Since execution packets only traverse in the up* direction of the spanning tree, each node must only store a single pointer per spanning tree (the gradient direction). An execution packet's ability to find the appropriate resources depends on several fabrication variables, including defect rates and the distribution of node types (evaluating this space is future work). The next section describes how to exploit the packet routing infrastructure to configure a fully addressable memory system in each cell.

7.7 Memory

Each cell represents a local namespace for memory and includes both data and instructions. The memory system must be able to (1) allocate (number or name) its locations, (2) provide an interface to execution packets, and (3) route memory packets (both requests to specified locations and responses back to requestors).

7.7.1 Memory Allocation

The memory network is a spanning tree rooted at the cell anchor. To configure memory, allocation packets are injected from the via through the anchor node, initially routed on virtual channel zero using any planar gradient. When an unallocated memory node receives an allocation packet, it records the address, marks itself as allocated, and sinks the packet. The second allocation packet received by this node is forwarded along the specified gradient, forming a branch in the network. For the third allocation packet, the node modifies the header to route the packet on virtual channel one along a planar gradient that creates a second branch in the network. Three-fourths of the subsequent allocation packets arriving on virtual channel zero are forwarded along the first branch, while the remaining packets use the other branch and switch to virtual channel one. Packets on virtual channel one are never modified. Cycles in the memory network are prevented by having an allocated node only accept configuration packets on the same physical link as its original allocation packets.

Memory ports are allocated after memory nodes and must have three good links (excluding the link used by the incoming packet) with three distinct planar gradients. Ports never change an allocation packet gradient, thus keeping the remaining two links free for the execution network. Memory ports are unnamed except for one port where execution is initiated. Non-memory nodes between memory nodes route allocation packets according to the specified gradient and reserve the corresponding links only for memory operations. A second planar gradient configuration creates new spanning trees that do not include any of the memory network links, thus creating two separate networks. Figure 7.7 illustrates the allocation of 64 memory locations

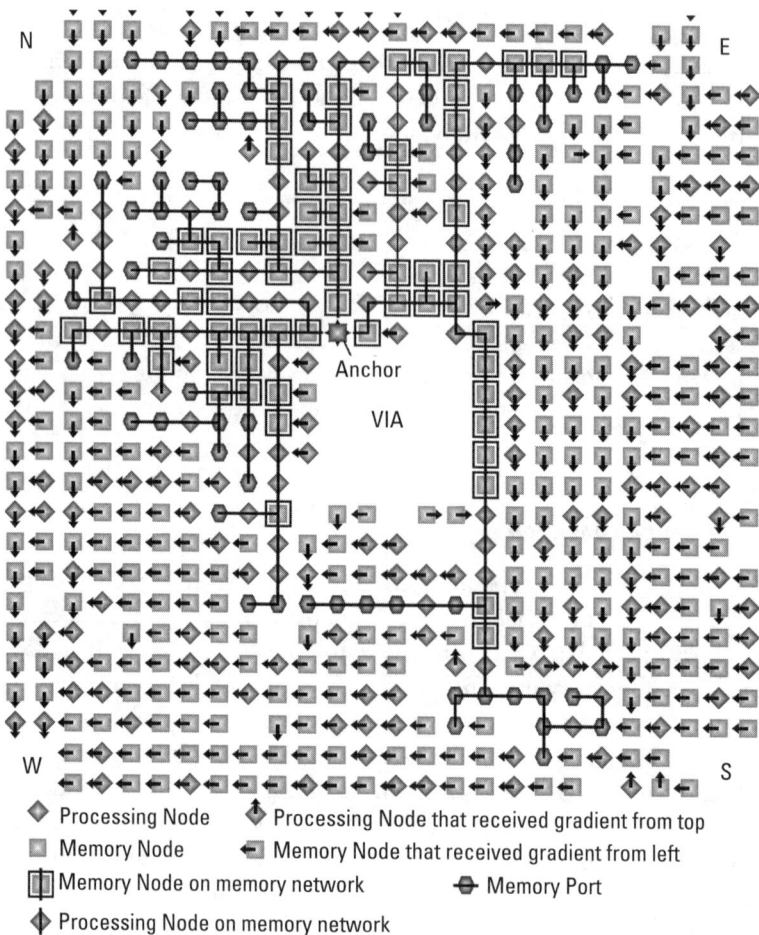

- ◆ Processing Node
- ■ Memory Node
- ▦ Memory Node on memory network
- ◆ Processing Node on memory network
- ◆ Processing Node that received gradient from top
- ◀ Memory Node that received gradient from left
- ◆ Memory Port

Figure 7.7 Memory network. 32 × 32 grid with a fully configured memory network, showing one gradient (west).

and 64 ports on a 32 × 32 grid with a 3% defect rate. For illustration only, the system includes only the west planar gradient on the execution network and uses a low defect rate on a small grid. Clearly, in this memory system the anchor node could be a bottleneck.

7.7.2 Interfacing Execution and Memory

The interface between the execution network and the memory network is controlled by memory ports that assume responsibility for handling all memory operations, including the JMP/CALL instructions for packet instantiation (see Section 7.8). When an execution packet needs to perform a constant or indirect memory operation, it searches for a memory port. A memory port servicing an execution packet stalls the execution packet, but at different points for loads and stores. Since load addresses are contained in the instruction field, the load can immediately issue and only stall the packet when the first bit of the operand stream arrives. Thus, the header continues searching for resources for subsequent instructions. When the memory port that initiated the load receives the response, it interleaves the memory contents into the execution packet's operand stream, enabling the operand stream to continue forward. A store must see the entire operand stream to extract the data, and after the node issues the store, it stalls the packet until the store is acknowledged. This acknowledgment ensures interpacket memory disambiguation. Memory ports also support indirect memory operations which require back-to-back memory operations: one to load the address and the other to access the contents at that address. This is implemented by first issuing a constant load to obtain the address, then using the result to generate another address for the load or store.

7.7.3 Routing Memory Packets

Memory packets are routed on either a request or response virtual network (two virtual channels per unidirectional link) that each obey up*/down* routing. Routing in the up direction follows the cell gradient up the spanning tree to the anchor node where the packet is broadcast in the down direction. Broadcasting is necessary since the destination memory node or port could be anywhere in the memory network. Loads require two full traversals of the memory network. However, since the anchor node is a serialization point for memory operations, it can acknowledge a store by broadcasting down the response network. Memory operations for addresses outside the originating cell are passed by the anchor onto the microscale network.

7.8 Packet Instantiation and Chaining

Entire execution packets (from header through tail) can be stored in memory by fragmenting them across memory nodes. Each fragment contains a portion of the execution packet and the memory address of the next sequential fragment (zero indicates termination). The fragments are written into memory using the microscale interface to inject store requests into the memory network. Packets are reassembled and instantiated on the execution network at a memory port using special sequencing instructions. Initial execution starts by using the microscale interface to inject one of these instructions on the memory response network for the named memory port.

Chaining is the process of sequencing instructions or packets under software control by including a special instruction as the last operation. The architecture supports two forms of the sequencing instruction: (1) CALL creates an entirely new packet, but stalls until all previous instructions are complete (i.e., it sees the packet tail); and (2) JMP injects new operations into the existing packet by stalling the operand stream, thus enabling accumulator forwarding. Conditional CALL is easily supported since the instruction waits for the packet tail. Execution of one packet can overlap with its dependent packet's search for functional and memory nodes. Full exploration of the instruction set and various forms of parallelism are interesting areas of future research.

7.9 Improving Node Utilization

While the four planar gradients allow routing execution packets in the cell, only a small fraction of all execution resources in a cell are used. This is because the route taken by the execution packet depends on its insertion point in the cell and the gradient that is being used to route. The execution network within the cell does not have a well-defined structure if the system uses planar gradients for routing. Improving the number of nodes reachable by execution packets requires modifying the structure of the execution network within a cell.

This is accomplished by adding a second via and anchor node ("execution anchor") to the cell. This via is used only by the execution network. Once the memory system has been created, an "execution" gradient is broadcast in the cell. This gradient reaches nodes that have not been included in the memory network and any ports on the memory network. This allows creation of an execution network with better structure by performing a depth-first traversal on the spanning tree created during the broadcast of the execution gradient. All execution packets follow this depth-first order ensuring high execution node utilization. The memory and execution networks now include most of the nodes in the cell, potentially allowing the use of about 97% of the cell (some nodes can become isolated during the creation of the memory network). However, as discussed in Section 7.11, there are other aspects of NANA that limit node utilization.

7.10 Preliminary Evaluation

This section presents a preliminary evaluation of NANA. The goal is to demonstrate the viability of the approach and to provide more details on execution. CNFET device characteristics suggest that this technology may have significant advantages over silicon in terms of power, delay, and cost. Collaborations with physical scientists to fabricate and characterize CNFET electronics will enable quantitative evaluation, especially when combined with a tool chain to support automated circuit design [20, 21] and architectural evaluation. An initial node floorplan and description of the simulation framework can be used to demonstrate system operation and provide preliminary (i.e., to actual fabrication) performance results using two simple programs: (1) Fibonacci is strictly an illustrative example; and (2) string matching reveals the potential to exploit massively parallel computation with nanoelectronics. The section concludes with an analysis of the strengths and weaknesses of the proposed design.

7.10.1 Node Floorplan

Figure 7.8 shows an initial floorplan for a 3×3-μm node that includes both an ALU and 16 bits of data storage with 8-bit addresses. The four semicircles

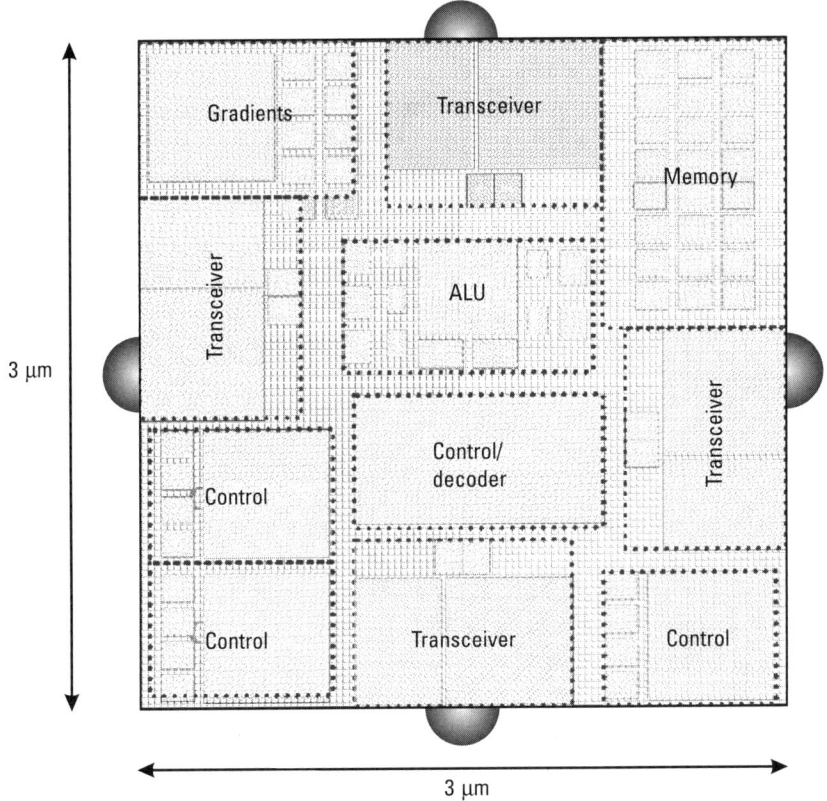

Figure 7.8 Node floorplan.

around the node represent contact points for internode links. The four transceivers control data transfer between the node and its neighbors. Configuration and gradient state is stored in the block denoted Gradients, while control logic is distributed in the four blocks marked Control, one of which is also responsible for decoding instructions (marked Control/Decoder). The small unlabeled blocks next to transceivers are the interface between the transceiver and the control/data logic of the node. The largest area is consumed by the various state machines sized according to the requirements derived from the simulator. The current implementation assumes specialized nodes, enabling more area for control and buffering.

7.10.2 Simulation Framework

NANA is evaluated using a custom event driven simulator written in C++ that simulates the system in detail. The simulator models activity within each node down to bit exchanges between components. The simulator models all node types and the system at all stages, including gradient broadcast, memory configuration, execution configuration, and run-time. It allows the user to vary a number of system parameters including the size of the network, node type distribution, event latencies, defect rate, and number of cells being simulated. Each cell holds a different part of the global address space and can execute different programs that are provided as input to the simulator. All events in the simulator are assumed to be a multiple of the clock cycle time (0.1 ns). The simulator accepts user-defined network topologies, or it can generate regular grid-based topologies. For simplicity, the evaluations assume a grid-based topology with a single 1,024 node cell and a 3% node defect rate in the evaluation. As long as the defect rate is low (about 15% or lower), the network topology has little effect on performance.

7.10.3 Fibonacci

This section considers the simple code that computes the Nth Fibonacci number. Table 7.4 shows the packet needed to implement Fibonacci for $N \geq 1$ (N is stored at address 0x30), and how the fragments are arranged in memory. For simplicity, each instruction is a separate fragment. The first packet, starting at address 0x10, loads the value N (counter), which specifies which Fibonacci number to compute, and the constants 1 and 0 (preloaded into 0x32 and 0x34 to begin with). The fourth instruction decrements the counter and sets the condition bit in the tail if the counter is zero. The counter is then stored back at address 0x30. The seventh instruction swaps the first two operands in the operand stream. The eighth instruction creates a copy of the accumulator at the end of the operand stream. The ninth instruction (ADD) computes the next Fibonacci number. If the condition flag in the tail is set, this new computed value is stored at address 0x36. The two remaining operands are then stored at locations 0x34 and 0x32. Finally, if the condition flag is not set, the code loops back to the beginning using a CALLZ instruction, creating a new packet. If the condition flag is set, the instruction is not executed, terminating the program. Figure 7.9 illustrates the creation of this packet with a bootstrapping JMP. Figure 7.9(a) shows the bootstrapping packet inserted at the via in the execution network. This packet is routed along the execution network until it finds a memory port. The JMP instruction in the packet executes at the port and starts fetching data from location 0x10 (where the Fibonacci code is stored). The data returned from location 0x10 [Figure 7.9(b)] is divided into two parts: (1) data for packet, and

Table 7.4
Packet Layout

Address	Op	Next	Address	Op	Next
0x10	LD (0x30)	0x14	0x26	CPACC	0x28
0x14	LD (0x32)	0x18	0x28	ADD	0x2A
0x18	LD (0x34)	0x1A	0x2A	CST (0x36)	0x2E
0x1A	DEC	0x1C	0x2E	ST (0x34)	0x40
0x1C	CMPZ	0x20	0x40	ST (0x32)	0x44
0x20	ST (0x30)	0x24	0x44	CALLZ (0x10)	0x0
0x24	SWAP	0x26			

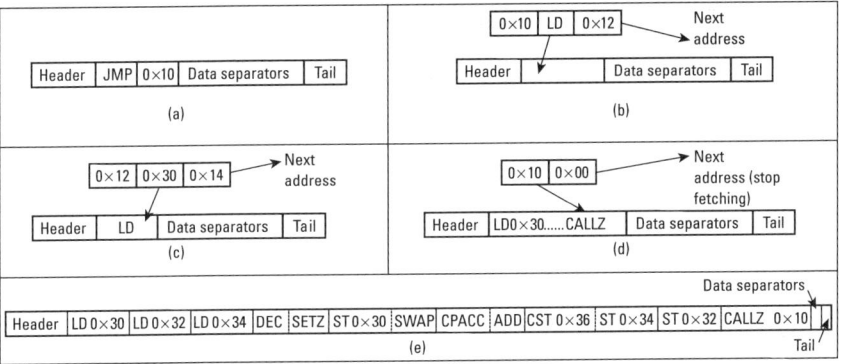

Figure 7.9 Bootstrapping the Fibonacci execution packet with a JMP.

(2) next address. The data for the packet (in this case, a LD opcode) is inserted into the packet and sent out on the execution network. The next address is used to fetch the next fragment of code (in this case, from address 0x12). The data returned from location 0x12 [Figure 7.9(c)] provides the address for the LD instruction and the address of the next fragment of code. This process is repeated until a data fragment with 0x00 as the next address [Figure 7.9(d)] is observed. This indicates that the JMP instruction is finished executing. The final packet before execution begins is shown in Figure 7.9(e). It is important to note that execution can begin while the JMP instruction is still executing.

To demonstrate system operation, its behavior is simulated at the bit-serial link level executing the above packets. A single 32 × 32 cell with 25% ALU nodes and four corner vias for planar gradients is modeled. A random distribution of defective nodes is assumed, with 3% of all nodes being defective. The memory system in the cell includes 64 16-bit memory nodes and 80 ports. A system using a depth-first execution network would achieve similar performance (depth-first execution only increases the number of nodes reachable on the execution network). The average time per loop iteration (13 instructions) is 22,300 cycles and it might be possible to reduce this through loop unrolling. However, only 2,000 of the 22,300 cycles are spent in performing the actual computation. More than 20,000 cycles are spent in accessing the memory system. Figure 7.10 illustrates the execution of the program. It is a snapshot of execution before the first load operation completes. While the absolute performance of this example does not surpass even current CMOS, it serves to demonstrate the operation of a single cell. The greatest advantage of this technology arises from the scale of the system.

7.10.4 String Match

The opportunity for massively parallel computation is tremendous. String searching is a common operation in many applications (e.g., searching for particular DNA sequences within a genome). The string match program loads a 16-bit key and compares it to all data elements within the cell, and a conditional store indicates if a match was found. This implementation requires 48 memory locations for instructions and 16 for data. Therefore, this system can search a 32-GB database by using all 10^9 cells. The execution time within one cell is 35 ns per comparison, for a total of 28.5×10^6 comparisons/sec. The potential for massive parallelism would be exposed by having each of the 10^9 cells perform a unique comparison, yielding an overall rate of 2.85×10^{16} comparisons/sec.

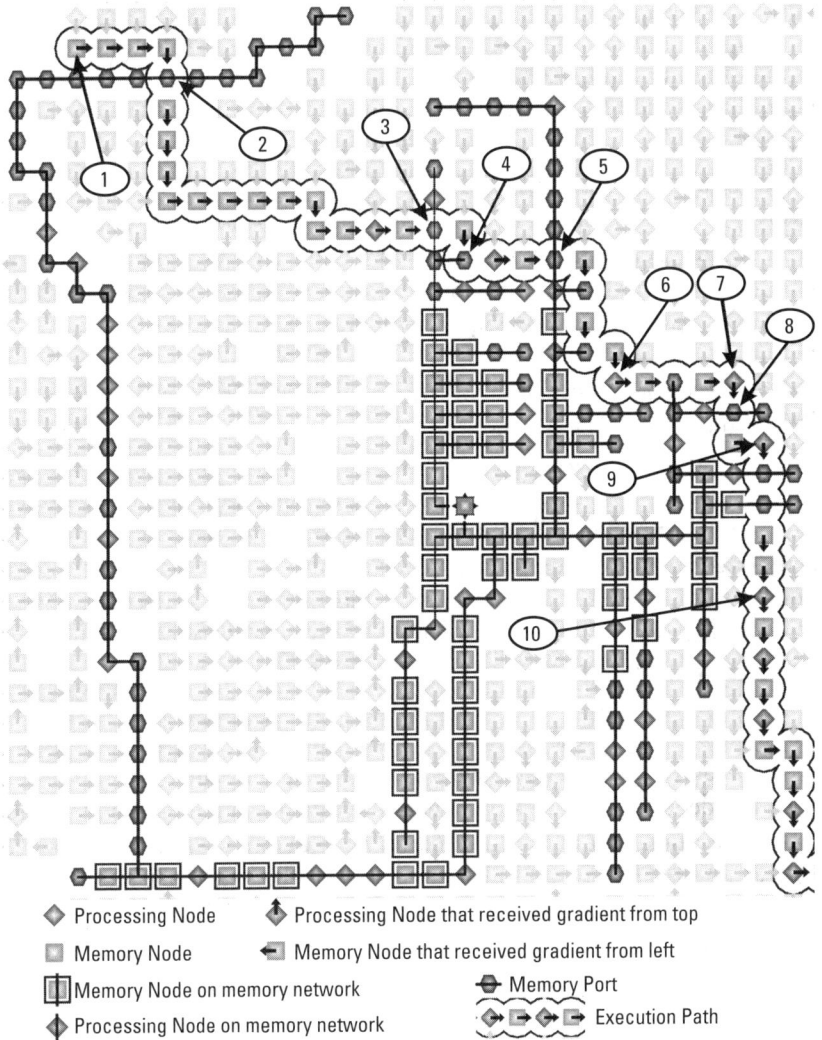

- ◇ Processing Node
- ◈ Processing Node that received gradient from top
- ▨ Memory Node
- ◀ Memory Node that received gradient from left
- ▥ Memory Node on memory network
- ⬢ Memory Port
- ◆ Processing Node on memory network
- ◇▶◆▶ Execution Path

Figure 7.10 The path of Fibonacci code in one direction through configured network with 1,024 nodes and 3% defects. Unused nodes in the execution network appear faded, defective nodes are omitted. 1: Bootstrap packet injected via; 2: JMP executes at port; 3: LD 0x30 executes at port; 4: LD 0x32 executes at port; 5: LD 0x34 executes at port; 6: DEC at processing node; 7: CMPZ executes at processing nodes; 8: ST 0x30 executes at port; 9: SWAP executes at processing node; 10: CPACC executes at processing node.

7.11 Discussion

The peak performance of NANA (assuming half of the nodes compute) is significantly higher than today's supercomputers. NANA can potentially perform 4.12×10^{21} bitops/sec, while the IBM Blue Gene can achieve 4.6×10^{16} bitops/sec and the NEC Earth Simulator can achieve a peak of 5.2×10^{15} bitops/sec. However, it will be a challenge for NANA to realize this peak performance in practice. Developing these programs exposes two key limitations of the current architecture: (1) underutilization of nodes and network connectivity, and (2) bottlenecked memory system.

Underutilization of Nodes. One of the key limiting factors to achieving good performance is the fact that nodes spend only a small fraction of their time doing useful work. For example, executing 10 arithmetic instructions, the node that executes the first instruction is doing useful work only when (1) it is receiving the first instruction and (2) it is receiving its operands for execution. Since there are 10 instructions to execute, which requires 11 operands (assuming data is preloaded), the packet will contain 868 bits (including header, instructions, operands, field separators and tail). Out of these, only 220 bits (header, instruction to be executed, separators, two operands, the operand separators and tail) are relevant to the execution of the instruction. Thus, the node is doing useful work only when it is dealing with ~25% of the bits in the packet. No useful computation is performed by the node in the remaining time.

The depth-first execution network increases the number of nodes usable during execution, but it does not reduce node idle time. The execution network can be thought of as a pipeline of nodes. The pipeline is most efficient only when it is full. Similarly, the execution network is fully utilized only when all nodes are actively executing instructions. This would require the creation of extremely long packets. However, the longer the packet, the longer it takes for a node to execute instructions because longer packets typically have longer instruction and data fields and a node needs to forward the entire packet before it can handle the next packet. This limits the peak performance of NANA.

Memory System Bottleneck. The memory system in NANA has multiple bottlenecks. Because of the way it is designed, it is currently not possible to execute store instructions (direct, indirect, or conditional) from a packet before any load instructions (direct or indirect). This limits the size and content of execution packets that can be created. In addition, all memory requests are serialized through the anchor node. This creates a substantial bottleneck at the anchor node. There is no easy way of alleviating this bottleneck, without significantly adding to the complexity of the system. Finally, the limited routing capability in the random network limits the ability to build a balanced memory network. This often results in unbalanced networks with long latencies. Despite its limitations, NANA demonstrates that it is possible to build a computing system despite the severe technological constraints. As a first step NANA does remarkably well. Future designs based on this technology can use the insights gained from this design (as does the architecture described in Chapter 8). NANA is a necessary first step toward exploiting nanotechnology's potential to overcome the red brick wall and provide architectures that more closely match conventional computing models.

7.12 Conclusions

This chapter presents an architecture that addresses the challenges posed by DNA-based self-assembly of carbon nanotubes and other nanotechnologies with similar characteristics (possibly even scaled CMOS). To overcome (1) limited node size, (2) random interconnection of nodes, and (3) a high defect rate, an active-network architecture with an accumulator-based ISA is presented. This architecture enables execution packets to search through a sea of heterogeneous nodes for the functionality they need, while avoiding defective nodes. An initial configuration phase imposes limited structure on the computing substrate, particularly for routing and memory allocation. Simulations of this architecture running simple programs demonstrate its viability, and provide preliminary performance numbers. While this architecture is only a relatively unoptimized first step, it addresses some of the key challenges in this class of nanotechnology and it highlights the technology's architectural implications.

References

[1] Tennenhouse, D. L., and D. J. Wetherall, "Towards an Active Network Architecture," *Computer Communication Review*, Vol. 26, 1996.

[2] Patwardhan, J. P., et al., "Evaluating the Connectivity of Self-Assembled Networks of Nanoscale Processing Elements," *Proc. IEEE International Workshop on Design and Test of Defect-Tolerant Nanoscale Architectures (NANOARCH '05)*, pp. 2.1–2.8.

[3] Culbertson, W. B., et al., "The Teramac Custom Computer: Extending the Limits with Defect Tolerance," *Proc. IEEE International Symposium on Defect and Fault Tolerance in VLSI Systems*, pp. 2–10.

[4] DeHon, A., "Array-Based Architecture for Fet-Based, Nanoscale Electronics," *IEEE Transactions on Nanotechnology*, Vol. 2, No. 1, 2003, pp. 23–32.

[5] Goldstein, S. C., and M. Budiu, "NanoFabrics: Spatial Computing Using Molecular Electronics," *Proc. 28th Annual International Symposium on Computer Architecture (ISCA)*, pp. 178–191.

[6] Heath, J. R., et al., "A Defect-Tolerant Computer Architecture: Opportunities for Nanotechnology," *Science*, Vol. 280, 1998, pp. 1716–1721.

[7] Berkel, K. V., and A. Bink, "Single-Track Handshake Signaling with Application to Micropipelines and Handshake Circuits," *Procceding of the Seconds International Symposium on Advanced Research in Asynchronous Circuits and Systems*, 1996, pp. 122–133.

[8] Kim, H.-S., and J. E. Smith, "An Instruction Set and Microarchitecture for Instruction Level Distributed Processing," *Proc. 29th Annual International Symposium on Computer Architecture*, pp. 71–81.

[9] Lavington, S. H., "The Manchester Mark I and Atlas: A Historical Perspective," *Commun. ACM*, Vol. 21, No. 1, 1978, pp. 4–12.

[10] Campbell-Kelly, M., "Programming the EDSAC: Early Programming Activity at the University of Cambridge," *IEEE Annals of the History of Computing*, Vol. 20, No. 4, 1998, pp. 46–67.

[11] Kim, H.-S., and J. E. Smith, "Dynamic Binary Translation for Accumulator-Oriented Architectures," *Proc. International Symposium on Code Generation and Optimization: Feedback-directed and Runtime Optimization*, pp. 25–35.

[12] Dally, W. J., "Virtual-Channel Flow Control," *Proc. 17th Annual International Symposium on Computer Architecture*, pp. 60–68.

[13] Dalal, Y. K., and R. M. Metcalfe, "Reverse Path Forwarding of Broadcast Packets," *Communications of the ACM*, Vol. 21, No. 12, 1978, pp. 1040–1048.

[14] Schroeder, M. D., et al., "Autonet: A High-speed, Self-Configuring Local Area Network Using Point to Point Links," *IEEE Journal on Selected Areas in Communications*, Vol. 9, 1991.

[15] Glass, C. J., and L. M. Ni, "The Turn Model for Adaptive Routing," *Proc. 19th Annual International Symposium on Computer Architecture*, pp. 278–287.

[16] Johnson, D. B., and D. A. Maltz, "Dynamic Source Routing in Ad Hoc Wireless Networks," *Mobile Computing*, Vol. 353, 1996.

[17] Intanagonwiwat, C., R. Govindan, and D. Estrin, "Directed Diffusion: A Scalable and Robust Communication Paradigm for Sensor Networks," *6th Annual International Conference on Mobile Computing and Networking*, 2000, pp. 56–67.

[18] Castro, M., and B. Liskov, "Practical Byzantine Fault Tolerance and Proactive Recovery," *ACM Trans. Comput. Syst.*, Vol. 20, No. 4, 2002, pp. 398–461.

[19] Dwyer, C., et al., "DNA Self-Assembled Parallel Computer Architectures," *Nanotechnology*, Vol. 15, 2004, pp. 1688–1694.

[20] Dwyer, C., M. Cheung, and D. J. Sorin, "Semi-Empirical SPICE Models for Carbon Nanotube FET Logic," *Proc. IEEE Conference on Nanotechnology*, pp. 386–388.

[21] Dwyer, C., et al., "Design Tools for a DNA-Guided Self-Assembling Carbon Nanotube Technology," *Nanotechnology*, Vol. 15, 2004, pp. 1240–1245.

8

A Self-Organizing Defect Tolerant SIMD Architecture

DNA-based fabrication produces precise control within a small area (e.g., 9 μm^2) enabling the construction of a large number ($\sim 10^9$ to 10^{12}) of small nodes (computational circuits with $\sim 10^4$ transistors) that can be linked together using self-assembly. This produces a random network of nodes, due to the lack of control over node placement and orientation, which contains defective nodes and links. While the work described in this chapter is motivated by DNA-based self-assembly, it is applicable to any technology with similar characteristics (e.g., scaled CMOS with high process variability, high defect rates, and point-to-point links between relatively small compute nodes). The common theme for all the architectures presented in this book is the challenge for computer architects to efficiently exploit the computational power of the large number of nodes while overcoming two primary challenges: (1) loss of precise control over the entire fabrication process, and (2) high defect rates.

This chapter presents a SIMD architecture designed to address these challenges, and it builds on the experiences gained through designing the previous three architectures. As with the previous architectures presented in this book, the fundamental building block in the architecture is a relatively small node (e.g., 1-bit ALU with 32 bits of storage and communication support for four neighbors) that operates asynchronously. A configuration phase at startup isolates defective nodes and allows groups of nodes to self-organize into SIMD processing elements (PEs) which are connected in a logical ring, thus simplifying the programmer's view of the system.

Simulations using conservative estimates for node size and device speed show that the proposed design can match the performance of aggressively scaled architectures for eight out of nine benchmarks tested. Furthermore, this performance is achieved with a very low power density of 6.5 W/cm^2 (versus power density greater than 75 W/cm^2 for modern cores) while conservatively assuming that about 90% of the devices in the system switch every nanosecond. Finally, the system can tolerate up to 30% defective nodes. These results demonstrate the potential of this technology for building high performance architectures despite high defect rates and loss of precise control during fabrication. Further improvements are possible as the technology scales to allow more complex nodes, better internode connectivity, and faster devices. The main contributions of this chapter are:

- Adapting self-organization methods to computer architectures;
- Designing a node that balances fabrication constraints with the functionality needed to communicate, compute, and self-organize;
- Demonstrating the above capabilities by composing a high performance, defect tolerant SIMD architecture from a random network of nodes.

8.1 DNA-Based Self-Assembled Nanoscale Systems and the Architectural Implications

Self-assembly of nanoelectronic devices has the potential to emerge as a lower cost alternative to top-down manufacturing. DNA-based self-assembly [1] uses the precise binding rules of DNA with nanoscale devices to build computing systems. Assume the previously described assembly process [2] to place electronic circuits on a DNA grid [3, 4]. The basic principle is to replicate a simple unit cell on a large scale to build a circuit. The unit cell consists of a transistor placed in the cavity of a DNA-lattice. A key requirement of this process is the ability to control the placement of electronic devices (e.g., carbon nanotubes [5, 6] or silicon nanowires [7]) at specific points on the DNA scaffold to form a circuit. Recently, two critical steps towards this goal were demonstrated: (1) aperiodic patterns, with a 20-nm pitch, on a DNA grid [8, 9]; and (2) DNA-guided self-assembly of nanowire transistors [10]. Figure 8.1 shows an atomic force microscope image of the letter A patterned on a DNA grid. Figure 8.2 shows a schematic of a small lattice with carbon nanotube-based transistors. Currently it is assumed there are only two layers of metal interconnect within a lattice, which limits the ability to place and route circuits. As described in Chapter 3, the fabrication approach proposes the use of conducting metallic planes separated by insulating layers to provide power and ground to the circuit. Figure 8.3 depicts a cross-sectional view of the lattice, with two layers of interconnect and the power and ground planes.

Current self-assembly processes produce limited size DNA grids and thus limit circuit size. However, the parallel nature of self-assembly enables constructing many nodes ($\sim 10^9$ to 10^{12}) that may be linked together by self-assembled conducting nanowires [4]. The proposed self-assembly method does not control the placement and orientation of nodes as they are intercon-

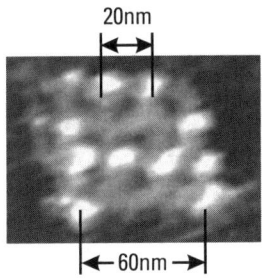

Figure 8.1 Patterned DNA. (*Source:* [8].)

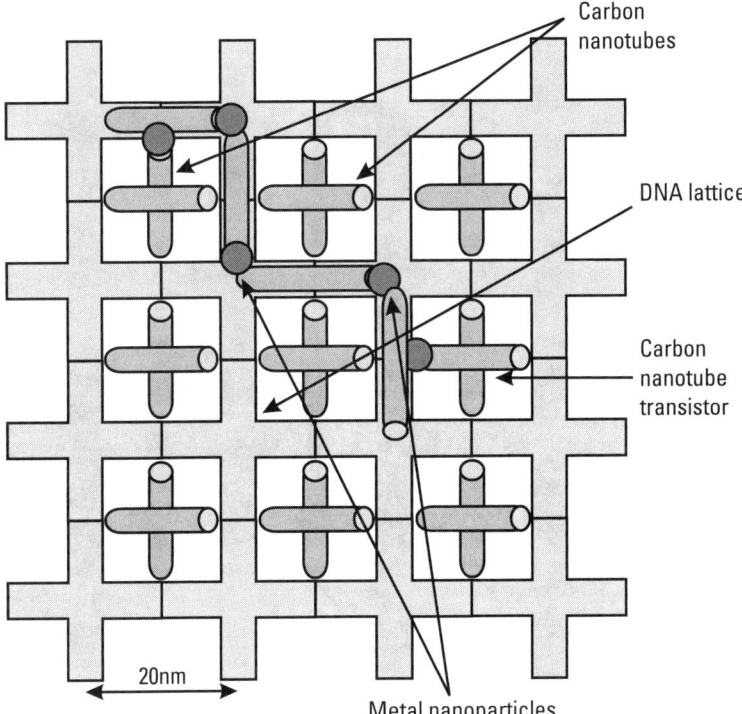

Figure 8.2 DNA Lattice with transistors and interconnect.

nected, resulting in a random network of nodes that contains defective nodes and links. Communication with external CMOS circuitry occurs through a metal junction (*via*) that overlaps several nodes but interfaces with the network of nodes through a single *anchor node*. There may be several via/anchor node pairs in large networks. Figure 8.4 shows a small network of nodes,

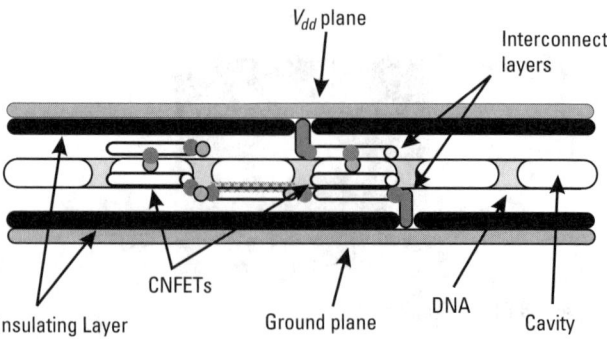

Figure 8.3 Lattice with two levels of interconnect and connections to Vdd and ground.

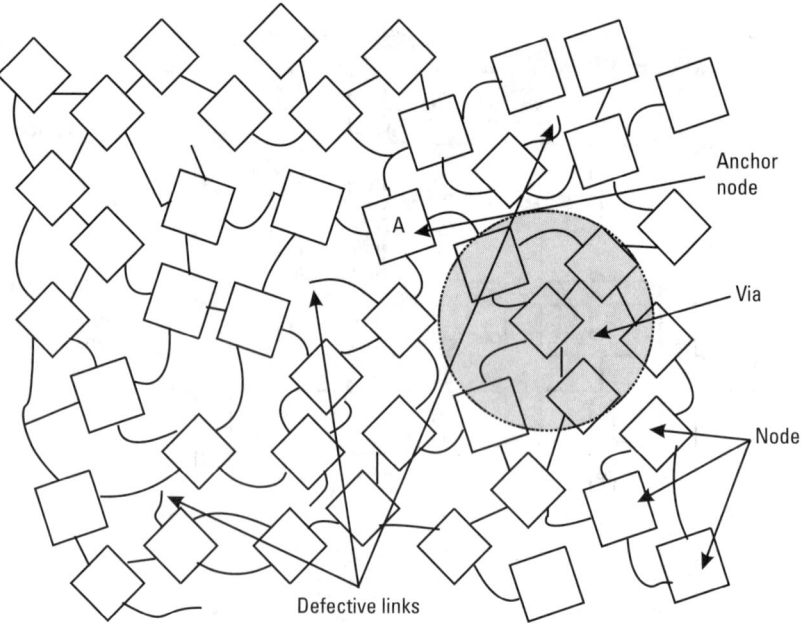

Figure 8.4 Self-assembled network of nodes.

including regions with defective links, and a via/anchor. In the rest of this chapter the term *anchor* refers to an anchor node/via pair.

A computing system built from this random network must: (1) tolerate node and interconnect defects, (2) not rely on underlying network structure, (3) compose more powerful computational blocks from simple nodes, (4) minimize communication overheads, and (5) achieve performance that is at least comparable to future CMOS-based systems. Several research projects examine building computing systems with a subset of these goals, including self-organization [11, 12], routing and resiliency in the face of defects [11, 13], and the ability to compose complex computational units from simpler blocks [14], but there are added challenges because of the extremely limited computational capabilities available in nodes. The nanoscale active network architecture (NANA) [15], presented in the previous chapter, is a general purpose architecture designed with a similar set of goals, assuming similar underlying technology. However, it fails to match the performance of conventional CMOS systems since it is unable to efficiently utilize the computational capabilities of the nodes at the same time. The design of the SIMD architecture presented in this chapter is guided by the lessons learned through the design and evaluation of NANA.

8.2 System Overview

The goal of the work presented in this chapter is to build a defect tolerant computing system with a random network of nodes using a mix of new solutions and adaptations of known techniques and to achieve performance comparable to future CMOS-based systems. A SIMD architecture, executing data parallel workloads, is used since it is capable of efficiently utilizing large numbers (greater than 10^9 to 10^{12}) of nodes. The proposed system—called the self-organizing SIMD architecture (SOSA)—supports a three-operand register-based ISA with predicated execution and explicit PE-Shift instructions to move data between PEs and communicate with an external controller that has access to a conventional memory system and that executes conditional branches, such as loops, that cannot be implemented with predication.

Each self-assembled node is a fully asynchronous circuit and there is no global clock to synchronize data transfers between or within nodes. Each node has a 1-bit ALU with a small register file and connects to other nodes with (up to four) single wire links. Each link supports low bandwidth asynchronous communication that transfers 1 data bit per handshake. To support deadlock-free routing, the design includes support for three virtual channels (1 bit each). The random network of nodes is organized at two levels during a configuration phase. First, since a node is too small to hold a PE, a set of nodes is grouped to form a PE. Second, PEs are linked in a logical ring providing programmers a simplified system view to reason about inter-PE communication.

The configuration process, initiated from an anchor, maps out defective nodes and connects functional nodes in a broadcast tree. The system can be configured in two ways: (1) as a monolithic system, all nodes on one logical ring (one cell); or (2) as multiple, independent logical rings (multiple cells). For a monolithic system, anchors can be used to speed up PE configuration and data input/output by serving as *taps* into the logical ring. The only constraint enforced during configuration is that an anchor cannot partition a PE. In the second configuration, the system is space partitioned by running the configuration algorithm from multiple anchors to create independent cells. Space partitioning is a common technique used in highly parallel systems to increase resource utilization by enabling the execution of multiple instances of one workload, or running multiple workloads. The space partitioning for the benchmarks is discussed in Section 8.5.6.

In the next three sections, SOSA is described in detail. Though the system is presented by a bottom-up view, the actual design process was iterative and involved several passes through node and system design, requiring a balance between size constraints and adding hardware optimizations to improve performance.

8.3 Node Microarchitecture

Careful node design is critical in maximizing system performance. Due to limited node size, designing the node architecture involves a trade-off between maximizing functionality (compute, communicate, and self-organize) and performance while minimizing circuit size. To avoid the area and power overhead of routing clock signals and to mitigate the effects of device parameter variation, instruction, execution, and sequencing within a node are asynchronous. The rest of this section describes the node microarchitecture, splitting the discussion into the data path (Section 8.3.1), control (Section 8.3.2), and internode communication (Section 8.3.3), highlighting the trade-offs between functionality, performance, and circuit size (Section 8.3.4).

8.3.1 Data Path

Each node has a simple data path that consists of a 1-bit ALU, a 32-bit register file, and a 1-bit data buffer that stores incoming and outgoing data. The register file and data buffer can act as sources and/or sinks for the ALU. The data buffer cannot be written to unless the current instruction is waiting for data, and once written, cannot be overwritten until the data is used by the ALU. All internal node communication occurs on dedicated point-to-point links. Where possible, the latency of moving a bit between two parts of the node is overlapped with other operations.

Nodes can be designed to partition the 32-bit register file into N-bit wide registers that require an N-bit ALU or repeated use of a single-bit ALU. For example, a 32-bit PE could be created with 32 1-bit registers, requiring 32 nodes for the PE, or with 16 2-bit registers, requiring 16 nodes to form the PE. Increasing register width increases the work done per instruction in a node, reduces the number of nodes required to form a PE, and reduces inter-PE communication overheads (since PE length reduces). However, for a fixed sized node, wider registers reduce the number of registers available to a programmer. Simulations reveal that 2-bit wide registers achieve the best trade-off in terms of maximizing the benefit of wider registers and the number of registers available to programmers (see Section 8.5.7.3). Results also show that program performance is not sensitive to ALU execution latencies shorter than the time taken to send/receive a bit between nodes (see Section 8.5.7.4).

8.3.2 Control

The control logic in the node can be divided into two parts. The first part (configuration logic) is used only during configuration and has control registers for defect testing/isolation (main control register), and PE configuration (PE control register). Figure 8.5 shows a floorplan of the node with the configuration logic demarcated by a dashed rectangle within the control and data block.

The second part is the run-time control logic used to decode and execute instructions. Design complexity is reduced by sacrificing latency and using microcoded control logic with each instruction divided into multiple microinstructions. The run-time control logic has three control registers to hold each of three microinstructions that comprise an instruction: (1) opcode, (2) register specifier, and (3) synchronization (sync). The sync microinstruction holds an optional counter value (repeat counter) to enable the repeated execution of one instruction and avoid broadcasting the same instruction consecutively. The register specifier includes fields that allow simple increment/decrement operations on register specifiers in conjunction with their reuse (for striding through registers). The design includes a shared circuit that is used to increment/decrement register specifiers and the repeat counter. Because of high instruction execution latencies, the increment/decrement operations can be overlapped with other operations, effectively hiding their latency.

All arriving microinstructions are first sent to an instruction buffer before they are moved to the control registers, creating a simple two-stage pipeline (buffer, execute). Each entry in the instruction buffer can hold all three microinstructions that form a full instruction. The instruction opcode is fully

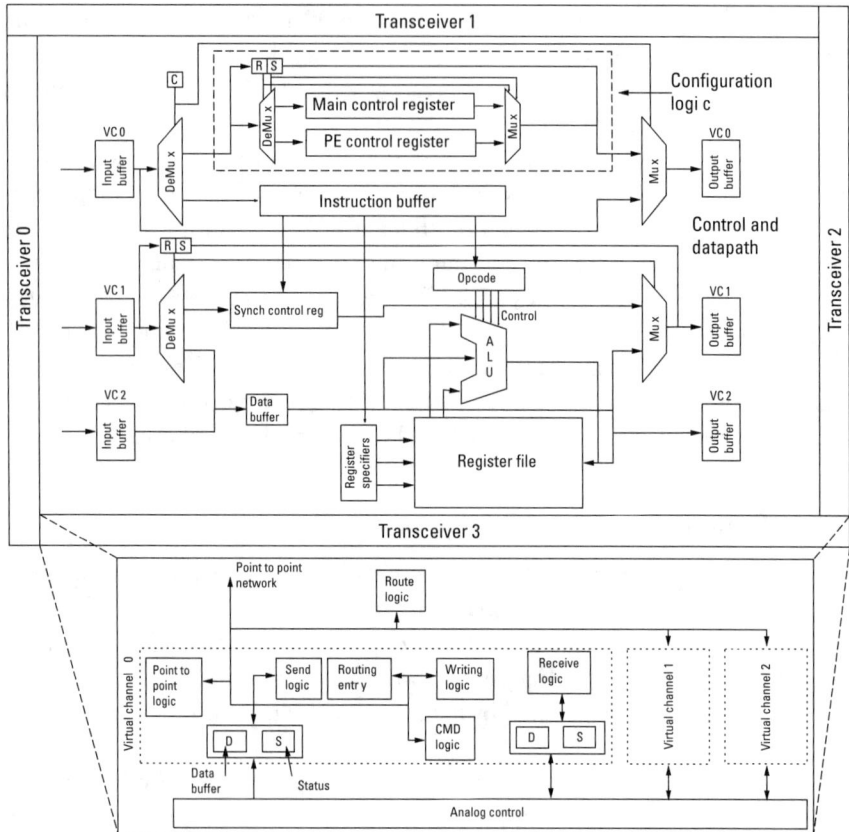

Figure 8.5 Node floorplan.

decoded, and copying the instruction into the control registers enables all control signals required to execute the instruction and detect its completion so that the next instruction can begin to execute. Increasing the instruction buffer size can improve performance by overlapping instruction broadcast with execution, but can also cause greater contention (and reduce performance) on the network since instructions and data must share link bandwidth. Simulations reveal that a single entry instruction buffer offers the best trade-off between improving performance and minimizing design complexity.

8.3.3 Internode Communication

Nodes communicate with each other on single-bit asynchronous links. Each end of a link terminates in a transceiver that can handle three virtual channels (using 1-bit buffers per virtual channel). The transceiver can route each virtual channel (VC) independently and requires 3 bits of state per VC to store the destination address. To support self-organization, nodes include logic to configure static routes (see Section 8.4.1). Virtual channel 0 (VC0) is used to broadcast instructions. Virtual channel 1 (VC1) and virtual channel 2 (VC2) are used to route data in opposite directions on the logical ring. Each asynchronous transaction on a link is controlled through a four-phase handshake. The links support bidirectional full-duplex transfers. To simplify transceiver circuit size and complexity, nodes transfer 1 bit per handshake (which severely limits link bandwidth).

8.3.4 Circuit Size and Power Estimates

There is a completed circuit design for all node components, except the transceivers. This design, in conjunction with layouts of simple logic blocks, enables an estimate for node size and power consumption. The simulator (discussed in Section 8.5.4) models the system in sufficient detail to make it relatively easy to extract a circuit model for most components. Figure 8.5 shows a floorplan of a node, showing the approximate position (not to scale) of the datapath, control, and transceivers. This enables an estimate that the entire node will require 10,000 transistors. Since the proposed fabrication technology currently imposes limitations on the number of metal layers, the final area of the node is estimated to be the equivalent of 22,000 transistors (based on experience in laying out circuits) which translates to a 3×3-μm node. Recent work [4] has shown that it should be possible to manufacture DNA grids of this size. The transistor overhead is large, but it enables support for defect tolerance.

Power consumption is estimated using the energy*delay product for carbon nanotube field effect transistor (CNFET) circuits [16]. An upper bound on the per-node power consumption is calculated assuming a switching speed of 1 ns (see Section 8.5.5) and estimated node gate and latch counts. During execution, the configuration logic and a large part of the register file are inactive (at most, three registers can be active). Accounting for these inactive elements yields a node activity factor of 0.88, which corresponds to a power consumption of 0.775 μW per node. To obtain an upper bound on the power density of this system, assume that nodes are packed with no space between them. Using the estimated node area (9 μm^2) and power (0.775 μW) yields a maximum power density of 6.5 W/cm^2, with a node activity factor of 0.88. As a point of comparison, this is much less than the power densities of current CMOS processors, which are greater than 75 W/cm^2. However, this comparison is between different technologies. Nonetheless, this estimate is pessimistic since the node activity factor is a conservative estimate, nodes cannot be packed perfectly, and defective nodes will further reduce power density.

8.3.5 Summary

Each node in SOSA is a small circuit that can communicate with up to four neighbors, store small amounts of state, and perform simple computation. To minimize area and power overheads, the nodes use asynchronous logic; however, like current processors, a significant area is consumed by control and communication circuitry. The challenge is to coordinate the operation of these nodes connected through an unstructured network to execute programs.

8.4 System Configuration

To use the random network of nodes to perform useful computation, a configuration mechanism imposes logical structure on the network and isolates defective nodes and links from the rest of the system. This allows nodes to self-organize and avoids the need for an external defect map, which would be impractical to obtain given the scale and bandwidth limitations of the system. Once defective nodes are isolated, the functional nodes are grouped to form PEs.

8.4.1 Logical Structure and Defect Isolation

Configuration uses a variant of the reverse path forwarding (RPF) algorithm [17, 18] to impose a logical tree structure on the network and isolate defects. When the system is powered up or reset, all nodes enter a configuration mode, steer incoming packets to the configuration control registers, and execute the distributed RPF algorithm. A small packet is inserted through an anchor and is broadcast on all its active links (the transceiver analog control circuitry tests the liveness of its physical link).

The RPF algorithm states that any node receiving the broadcast propagates it on all links except the receiving link if and only if the node has not seen the broadcast before. The node also stores the direction (gradient) from which it received the broadcast and sets up internal routing information based on this direction. Following the gradient through a set of nodes leads to the broadcast source—the tree root. A depth-first traversal is established by nodes locally selecting links in a predefined order relative to their gradient link. Opposite orderings are used for forward (VC1) and reverse (VC2) traversals. This method can be used to have all nodes in the system self-organize into a tree or it can be used to create multiple trees by initiating the broadcast through multiple anchors. For example, self-assembly of the random network of nodes could occur on a silicon wafer with a grid of vias (created using conventional lithography) to create a system with multiple anchors.

Defect isolation is achieved by (1) augmenting each node with built-in self-test to implement fail-stop behavior [19], and (2) including a simple test vector in each broadcast packet that each node must successfully execute before propagating the broadcast. Nodes failing the test are isolated since there is no path through the node. Simulations show that the gradient can reach a very large fraction of functional nodes (i.e., achieve good coverage) for node defect rates up to 30%. Handling more complex defects like Byzantine failures is beyond the scope of this work.

8.4.2 Configuring Processing Elements

A node is too small to hold an entire PE, so a set of nodes are logically grouped to form a PE. To create PEs with N bits (assume $N = 32$), the configuration packet traverses the broadcast tree in depth-first order (on VC1) and groups $N + 2$ consecutive, unconfigured nodes. There is one configuration packet per PE. An unconfigured node receiving a configuration packet examines it to determine what node in the PE is to be configured next. The first node holds auxiliary control bits for the PE and is called the head node. The next N nodes serve as compute nodes that form the N-bit PE. The last node (tail) serves as the terminating point of the PE and is used to store the status bits (carry/borrow) resulting from an arithmetic operation. A newly configured tail node sinks the configuration packet. To minimize PE setup time in large networks (greater than 10^9 nodes), configuration could be parallelized by exploiting multiple anchors.

If the broadcast tree does not have sufficient nodes to form an integral number of PEs, the incomplete PE is deconfigured before execution begins by performing a reverse depth-first traversal on VC2. PE deconfiguration uses a simple packet and starts with the last configured node of the partial PE (i.e., PEs with no tail), and deconfigures all intermediate nodes until it reaches (and terminates at) the head node. Figure 8.6 shows the logical order of nodes within a PE. Figure 8.7-(1) shows the network from Figure 8.4 in a configured state with three 8-bit PEs ordered by the depth first traversal of

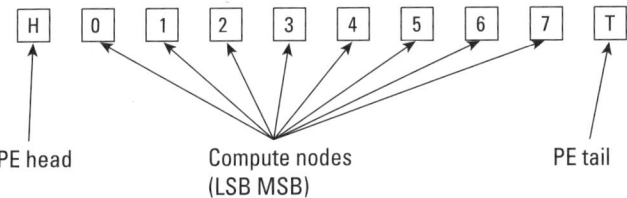

Figure 8.6 PE layout.

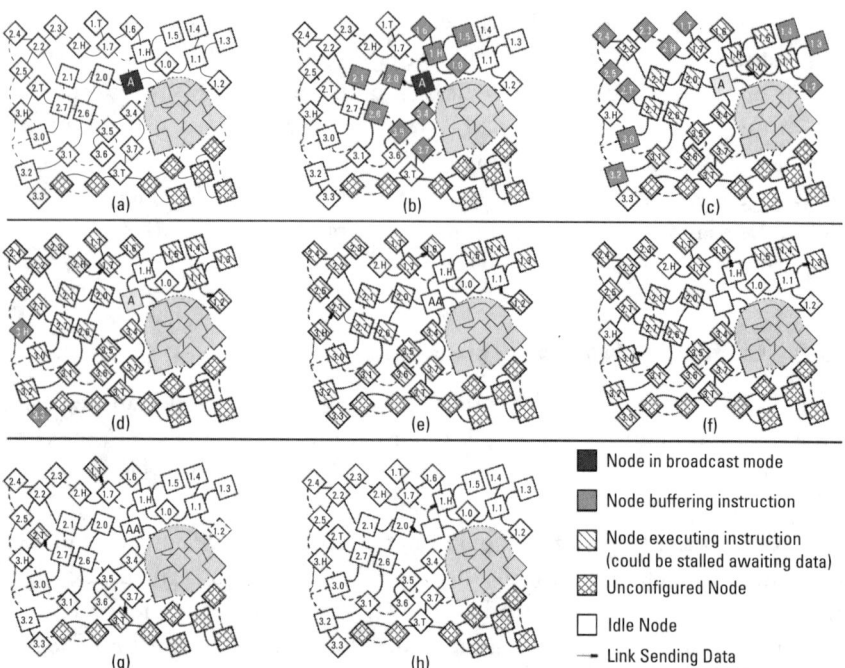

Figure 8.7 Instruction execution in a random network with three configured PEs. The via is shown to cover multiple nodes, which are rendered unusable. The via is connected to the PEs through the anchor node (A).

the network. The links shown with solid lines correspond to edges on the broadcast tree. Links that do not lie on the broadcast tree (dashed lines) are not used. The unlabeled nodes outside the via are part of a partial PE that has been deconfigured. The numbers within each node identify the PE that the node belongs to (first label) and the position of that node within the PE (second label).

Performance is enhanced by extending PE configuration to optimize PE length (hops from head to tail). Very long PEs (e.g., a PE that spans the broadcast tree root) may reduce performance due to longer intra-PE communication latencies. Since the postconfiguration step deconfigures partial PEs, a PE that crosses a length threshold can be rejected by starting a new PE without creating a tail node. By varying the PE length threshold, and determining the corresponding effect on performance, it is empirically determined that a threshold of four times the minimum PE length (compute nodes + head + tail) achieves a good balance between extra nodes required and performance gained by reducing PE length. Section 8.5.7 provides further details on this evaluation.

Once PEs are configured, all nodes set a run mode bit. Packets are no longer routed to the configuration control registers, unless the node receives a global reset instruction. Each PE waits for instructions to execute. The next section describes how SOSA uses the configured PEs to execute instructions.

8.5 System Architecture

Careful node design coupled with the self-organizing capability of each node enables mapping a data parallel architecture onto the random network of nodes. The instruction set is first described (Section 8.5.1) and then the execution model (Section 8.5.2). An example follows that illustrates the execution of an instruction in the system (Section 8.5.3).

8.5.1 Instruction Set Architecture

SOSA uses a three-register operand ISA, with microcoded instructions (Table 8.1 shows the instruction set). A full instruction has between 39 and 44 bits and contains: (1) a 16-bit fully-decoded opcode microinstruction, (2) a 20-bit register specifier microinstruction (4 bits per register specifier for a 16-entry register file, and 2 extra bits per register specifier to allow increment/decrement/no change operations), and (3) a 3-bit sync microinstruction with an optional 5-bit sync repeat counter. Each microinstruction can be independently broadcast and includes 2 bits of control overhead to select a control register as a destination. Since opcodes are fully decoded, it is relatively straightforward to support fused instructions that include combinations of operations to increase the work done per instruction. For example, a Copy-Shift first copies the source to the destination register, and then performs a shift operation on the destination register. SOSA also supports predicated instruction execution (all instructions can be predicated) and has three types of instructions that can modify predicate bits: (1) conditional instructions; (2) unconditional predicate modifying instructions; and (3) predicate-shift instructions.

Data exchange with the external controller and between PEs is handled through PE-Shift instructions. When PEs in a cell execute a PE-Shift instruction, each PE sends the contents of the specified register to a neighbor (left or right), and receives a new value for the register from the other neighbor (right or left). Since these instructions are critical for data communication, it is important to minimize their latency. PE-Shifts are optimized using the following observation: for a N-bit PE, each bit moves exactly $(N+2)$ positions to the left or right, and a node only needs to store the $(N+2)$th bit in its register file and can forward the remaining bits without register access. The optimization uses the sync repeat counter to track the bits being forwarded by the node.

Table 8.1
Instruction Set

Instruction Type	Opcodes	Description
Arithmetic	ADD, SUB, INC, DEC, SETGT, SETLT, SETEQ, SETNEQ	Various arithmetic and conditional instructions, Set instructions set the specified predicate register if the condition is satisfied
Logical	AND, XOR, OR, NOT	Various logical instructions
Shift	SHIFTML, SHIFTLM, PSHIFTML, PSHIFTLM	Various SHIFT instructions. ML = >MSB to LSB, LM = >LSB to MSB. The prefix P indicates that the instruction modifies the specified predicate register (not a predicated instruction).
PE-Shift	SHIFTMLPE, SHIFTLMPE,	PE-Shift instructions. Move register to adjacent PE.
Bit PE-Shift	BITSHIFTMLPE, BITSHIFTLMPE, MVSTCURRPE, MVSTNEXTPE	Shift single bits between PEs.
Register Operations	CLEAR, CPREG, SWAP	Clear, Copy, or Swap registers.
Predicated	PR[OPCODE]	Any instruction with the prefix PR is predicated. The predicate register corresponds to the first source register.
Predicate Modifying	PSet, PSetEven, PSetOdd, PInv	Predicate modifying instructions
Fused	CPSHIFTML, CPSHIFTM	Copies source into destination, and performs a shift on the destination
Signal	SIG_CTRL	Send signal to external controller

The node stops forwarding when it receives the $(N + 2)$th bit. When a node is forwarding data, it copies the data bit directly from its input buffer to its output buffer. This reduces the critical path of a bit through the node.

8.5.2 Execution Model

Instructions are broadcast on VC0 to all nodes, thus PEs, in a cell. Nodes first place instructions in the instruction buffer and then forward them down the broadcast tree. Instruction broadcast stalls when the instruction buffer is full. The arrival of the synchronization microinstruction is a signal to the node that all parts of the instruction have been received. An instruction moves from the instruction buffer to the node's internal control registers only when the previous instruction finishes execution. Since nodes are bandwidth limited, an optimization that utilizes partial broadcast of instructions reduces the number of bits broadcast. If an instruction broadcast skips a microinstruction (except sync), the node reuses the previously latched value from the corresponding control register. The sync repeat counter also helps reduce the number of bits broadcast.

Nonpredicated instructions can be executed independently by nodes of a PE, if there are no interbit data dependencies (e.g., for an OR instruction). The head and tail nodes act as PE delimiters, and ensure that intra-PE data packets do not cross PE boundaries. The tail node also stores the carry/borrow out from arithmetic operations. The head node stores predicate bits (one per physical register) that are used to conditionally execute predicated instructions. The head node reads the specified predicate bit and informs the remaining nodes in the PE whether the predicated instruction is to be executed or squashed by sending a sync microinstruction on VC1. Since each node in a PE must wait for the extra synchronization microinstruction (which is consumed by the tail), execution of predicated instructions is serialized through a PE.

8.5.3 Instruction Execution Example

Figure 8.7 uses the small configured network with three 8-bit PEs to illustrate the different steps involved in executing an ADD instruction. The anchor node broadcasts three microinstructions that form the ADD on VC0 (step 1). As each node receives the microinstructions it buffers them (step 2) and waits for the synchronization microinstruction to arrive. Once this microinstruction arrives (step 3), the node can start execution. Since the instruction to execute is an ADD, the head node of each PE must insert a carry-in for the first node (step 3). Each node then performs the ADD as it receives the carry-in (steps 4, 5, 6), and sends the carry-out to the next neighbor. When a node finishes with the ADD, it clears any temporary internal state used by the instruction and goes back to waiting for instructions to arrive (steps 7, 8).

One important aspect of the execution model is that different nodes and PEs can be in different stages of execution at the same time. In step 3, nodes 3.H and 3.3 are still idle, while other nodes in PE-3 are receiving data (3.0, 3.2), and some have received the full instruction and are stalled waiting for data (3.1, 3.4-3.T). This asynchronous execution within and between PEs allows them to make forward progress independently (as long as data dependencies are satisfied) and helps SOSA tolerate large internode communication latencies and achieve good performance.

8.5.4 Evaluation

This section describes the evaluation methodology, simulation infrastructure and workloads (Section 8.5.5), then compares SOSA performance to four other architectures (Section 8.5.6). The results show that SOSA achieves good performance on benchmarks that have data parallelism. For a configuration with more than 64K PEs, SOSA matches the performance of an ideal 16-way CMP. Thus, despite SOSA's severe limits on node computational power, network bandwidth and connectivity, and low control over the fabrication process, it matches the performance of idealized conventional architectures, with lower device switching speeds and a lower power density. Section 8.5.7 explores the sensitivity of SOSA to changes in various design parameters. The results also show that the instruction buffer and microinstruction reuse optimizations improve performance. Increasing ALU execution latency does not impact performance provided the total execution latency is less than communication latencies. Through redundancy SOSA can tolerate high node defect rates (Section 8.5.8). For the encryption benchmarks, performance gracefully degrades as the fraction of defective nodes increases to 30%. For the other benchmarks, by overprovisioning the system, SOSA tolerates up to 20% defective nodes with a small (<10%) degradation in performance.

8.5.5 Methodology

SOSA is evaluated using a custom, event-driven simulator which uses results from simulating smaller systems to extrapolate the behavior of larger systems. Since the nodes do not use a clock, the simulator defines the time taken to perform one part of the internode asynchronous communication handshake as one "time quantum." The latency of all activity in the node is a multiple of this time quantum. Experimental devices are expected to operate at frequencies exceeding 100 GHz [20] with demonstrated frequencies over 10 GHz [21] (time quantum of 0.1 ns), and asynchronous handshakes at high speeds have been demonstrated for high bandwidth crossbar networks [22]. SOSA's expected performance should scale with device performance, but assume a conservative time quantum of 1 ns to avoid overestimating performance due to aggressive technological parameters. The default simulation parameters are listed in Table 8.2. A custom tool models the growth of DNA nanotubes between nodes to generate network topologies.

The performance of SOSA is compared to a Pentium 4 (P4) (3 GHz, 1 MB L2, 1 GB RAM), an ideal out-of-order superscalar (I-SS) (128-wide, 8K ROB, 1-cycle memory latency), an ideal 16-way CMP (16-CMP) (obtained by linearly scaling performance of the I-SS), and an ideal implementation of SOSA (I-SOSA) that uses the same instruction set, but assumes unit instruction execution latencies, and no communication overhead. Table 8.3 lists the parameters used to simulate the I-SS with SimpleScalar [23].

Table 8.4 contains brief descriptions of the test programs, the broad application classes they fall under, and the number of PEs required by SOSA to run one instance of a program. For all programs other than the encryption algorithms, the system is configured as a single cell with the necessary PEs. For the encryption algorithms, the system is configured as a collection of cells, each of which operates as a pipelined encryption unit. PISA binaries for simplescalar are generated using gcc (flags: -O3) and Intel's C Compiler (icc, flags: -O3 -fast -tpp7) is used to generate binaries for the P4 since optimized icc binaries outperform optimized gcc binaries. Several versions of matrix multiplication [24] were evaluated and the best version selected for the P4 (naïve version with three nested loops, since icc vectorizes loops for the SSE units) and I-SS (static loop unrolling). Quicksort is the algorithm used for sorting. For SOSA each program is hand-optimized (e.g., loop unrolling, code reorganization). The SOSA code for matrix multiplication and the image filters

Table 8.2
SOSA System Parameters

Parameter	Value	Parameter	Value	Parameter	Value
Register file	16 entry, 2 bits/node	Sync repeat counter width	5 bits	Data width	32 bits
Time quantum	1 ns	PE length optimization	Enabled	Instruction buffer size	1 entry
ALU latency	1 time quantum	Register increment/ decrement	Enabled	Link type	Full-duplex

Table 8.3
Ideal Superscalar Parameters

Parameter	Value	Parameter	Value	Parameter	Value
Width	128 (fetch/decode/ issue/commit)	Integer ALU	128 ADD, 128 Mul	Branch prediction	Perfect
Instruction fetch queue	1,024 entries	FP ALU	128 ADD, 128 Mul	Memory latency	1 cycle
ROB/LSQ	8,192 entries, 1 cycle latency	Frequency	10 GHz	Memory ports	128

Table 8.4
Benchmark Descriptions

Application Class	Benchmark Description
Scientific	Multiply integer $N \times N$ matrices (N^2 PEs)
Image processing	Apply a generic 3×3 filter on an $N \times N$ image (N^2 PEs)
	Apply a separable Gaussian filter on an $N \times N$ image (N^2 PEs)
	Apply a median filter on an $N \times N$ image to reduce noise (N^2 PEs)
General purpose	Odd-Even Transposition Sort [25]: parallel sort with nearest neighbor communication (N PEs for sorting N numbers)
Cryptography	Tiny Encryption Algorithm (TEA): simple encryption algorithm used in the XBox (64 PEs)
	eXtended TEA (XTEA): eliminates known vulnerabilities in TEA (64 PEs)
Search	Search a database for a match with an input 32-bit string ($O(N)$ PEs for N strings)
Bin-packing	Pipelined version of bin-packing with first-fit heuristic (N PEs for N bins)

assumes data is in place before execution begins. However, this overhead forms only a small fraction of total execution time and can be reduced by exploiting multiple anchors in the system. The other workloads explicitly account for I/O overheads. The running times of programs do not include system configuration time (which is proportional to the number of nodes in the system). Simple extrapolation is used to estimate SOSA performance for configurations with more than 16K PEs (simulating a 256×256 matrix multiplication on a 3-GHz P4 with 32-GB RAM takes ~50 days, which is impractical for data collection purposes). For matrix multiplication, running time is extrapolated using $R(N) = 3.8*R(N/2)$, where $R(N)$ is the running time for an $N \times N$ matrix. For generic and median filters $R(N) = 1.85R(N/2)$, and for the separable Gaussian filter $R(N) = 2.77R(N/2)$. To validate the extrapolations, the extrapolated run-times are compared to simulated run-times for large configurations (8K to 16K PEs). Extrapolation is not necessary for sorting, since it is possible to simulate systems with up to 16K PEs (and hence, sort 16K numbers). The other workloads are throughput oriented, and do not require extrapolation.

8.5.6 Results

The following examines the performance of applications on SOSA with no defects. SOSA provides users the flexibility to configure the system to minimize program running time (single cell, single program instance), or to maximize throughput (multiple cells, one program instance each). The evaluation is divided in two parts based on the performance metric being used (execution time or throughput).

8.5.6.1 Execution Time

For many workloads (image filters, matrix multiplication, sorting), system performance is determined by program execution time since the user is interested in solving a single instance of each problem. To evaluate the performance of these programs on SOSA, the system is configured to create one cell with the required number of PEs. The latency of an individual instruction in SOSA is high due to the overheads caused by limited node capabilities. However, SOSA can amortize this overhead by executing the same instruction in all PEs at the same time. Hence, SOSA is expected to perform poorly for small input sizes, where each instruction is executed in a small number of PEs. However, SOSA performance should improve as input size increases and eventually match (or exceed) the performance of the P4, I-SS, and 16-CMP. The input size at which SOSA outperforms a particular architecture is application dependent.

Inspecting the main loop body for matrix multiplication in Figure 8.8 (optimizations are omitted to keep the code compact and readable; see the

```
; Initialize Before Multiply
CPREG R4, R2        ; Copy R4 → R2
CPREG R3, R1        ; Copy R3 → R1
CLEAR R5            ; Clear R5
; Multiply (Loop Dw times) (Dw: Data Width)
SHIFTLM R1          ; Shift LSB → MSB (multiply by 2)
PSHIFTML R2, R5     ; Shift MSB → LSB, LSB → Predicate register 5
PRADD R5, R1, R5    ; If predicate register 5 is set, R5=R1+R5
CLEAR R6            ; Clear R6
; Accumulate Partial Products
; Repeat log2(N) times (i is iteration count)
ADD R6, R6, R5      ; Accumulate Partial Sum
CPREG R6, R5        ; Copy R6 → R5
SHIFTMLPE R5        ; Repeat i*2 times
; End RepeatADD R6, R6, R5 ; Final Add
; Align rows of matrix A for next set
; of multiplies (Repeat N times)
SHIFTMLPE R4        ; Move 'A' N PEs to the left
; Move Result
CPREG R8, R9        ; If Pred. Reg 8 ==1, this PE holds the
                    ; first row/col element, move to R9
PSHIFTML R9, R6     ; Move that bit into the predicate register 6
PRCPREG R6, R7      ; if predicate register 6==1, copy R6 → R7
SHIFTMLPE R7        ; Move R7 one PE to the left
```

Figure 8.8 Matrix multiply: assembly code (no unrolling).

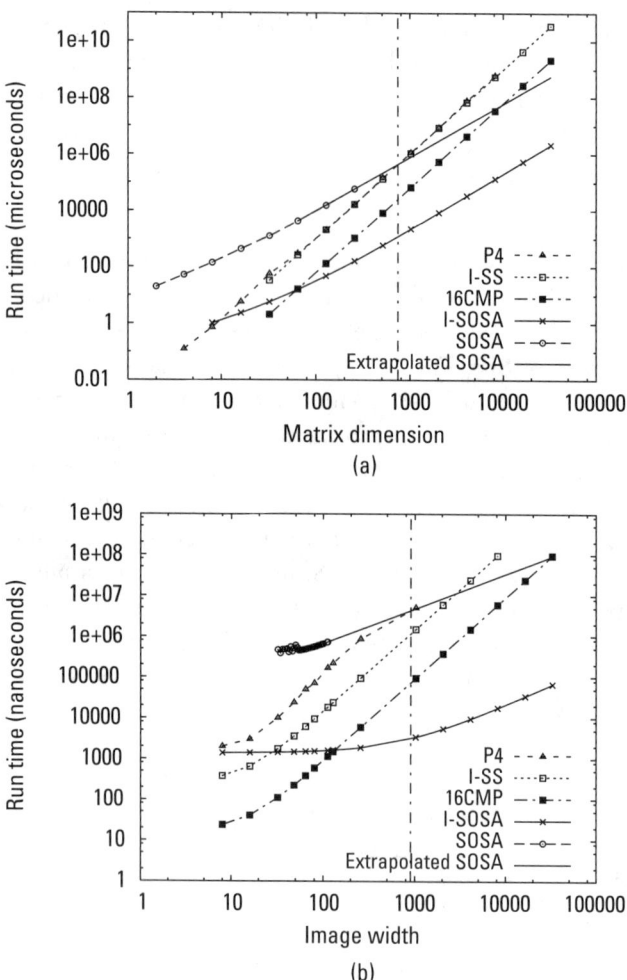

Figure 8.9 Single cell program run-times: (a) matrix multiplication, (b) Gaussian filter, (c) median filter, and (d) sort. The vertical line denotes the input size beyond which SOSA does better than the Pentium 4.

appendix for details on the optimizations performed to tune matrix multiplication) reveals that the primary advantage for SOSA is the simultaneous computation of all products in the N^2 PEs. This allows SOSA to convert the $O(N^3)$ algorithm to $O(N^2)$. Image filters and sorting are reduced from $O(N^2)$ algorithms to $O(N)$.

The plots of Figure 8.9 show the running time of matrix multiplication, Gaussian filters, median filters and sorting on different architectures, marking the input size beyond which SOSA outperforms the P4 with a verti-

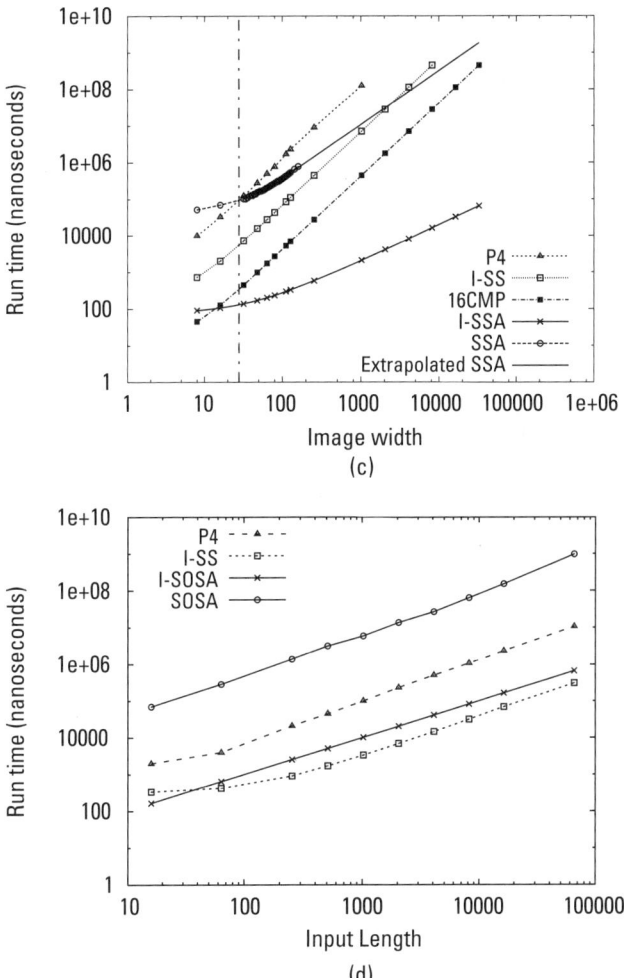

Figure 8.9 Continued

cal line (results for the generic 3 × 3 filter are qualitatively similar to the Gaussian filter). As expected, SOSA performs worse than the conventional architectures for small input sizes, but matches and overtakes them as input size increases (except for median filter and sort). The P4 matches the I-SS on matrix multiplication for two reasons: (1) the P4 makes use of its SSE units, and (2) I-SS only achieves an IPC of 9. The P4 performs much worse without the SSE units.

The performance of the median filter and sort algorithms is limited by their dependence on predicated instructions which serialize execution in a PE. While the number of predicated instructions in the median filter is fixed (independent of input size), for sort it scales with input size. For the median filter, SOSA is able to match the performance of the uniprocessors, but not the ideal 16-CMP (for image sizes up to 16K × 16K). For sort, the potential speedup on SOSA over quicksort on a single processor (average case) is $O(\log(N))$. However, the overhead introduced by predicated instructions makes it impossible for SOSA to match the performance of the I-SS or P4. Exploring techniques to reduce this overhead is future work. Note that even I-SOSA cannot outperform the I-SS at sorting. This highlights one key limitation of SOSA: it is not a general purpose architecture and cannot match the performance of conventional processors on general purpose workloads.

8.5.6.2 Throughput

There are a large number of workloads where high system throughput is desirable. The parallel computational capabilities of SOSA can be used to achieve high system throughput by dividing the system into multiple cells, each having a set of PEs. While there are multiple ways to improve throughput, the focus is on using multiple instances of a single application (operating on different data) running on different cells. For example, assuming an area of 100 mm^2 (approximately the area of a P4 in 90-nm CMOS) the system can configure more than 5,000 cells (assuming an average internode gap of 1 μm) that each perform an 8 × 8 matrix multiplication and achieve much higher throughput than the P4 or the I-SS.

TEA [26] and XTEA [27] are two simple encryption algorithms developed at the University of Cambridge that use a combination of shift, add, and xor operations to encrypt 64-bit blocks of data with a 128-bit key, with XTEA requiring more operations per iteration to achieve better cryptographic security. The software implementations are pipelined versions of both algorithms that require 64 PEs (corresponding to 64 encryption iterations) in a cell (i.e., 1 cell = 64 PEs). Due to their requirement of fixed sized cells, these algorithms are well suited for the high-throughput, multiple cell configuration.

Since each cell operates independently and can handle multiple data blocks in parallel, TEA and XTEA achieve better throughput on SOSA than on the I-SS or P4. A single cell can perform 175,000 TEA encryptions per second and 170,000 XTEA encryptions per second. Table 8.5 compares the performance of TEA on different architectures. The table shows that SOSA can achieve 79% of the throughput of the ideal 16-CMP, while using about the same area as a single core with devices switching at a 10th of the speed (1 ns versus 0.1 ns). The comparison with I-SOSA highlights the overheads due to simple nodes and limited bandwidth in SOSA.

Pipelined versions of searching and bin-packing algorithms are also implemented on SOSA to maximize throughput. The implementation of search

Table 8.5
TEA Throughput for Different Architectures

Architecture	Encryptions/sec
P4 at 3GHz (100 mm²)	3.9 M/second
I-SS	73.62 M/second
16-CMP	1,180 M/second
SOSA (1 cell ~ 0.019 mm²)	0.175 M/second
I-SOSA (1 cell = 64 PEs)	27.7 M/second
SOSA (5,400 cells ~ 100 mm²)	940 M/second
I-SOSA (5,400 cells)	72,300 M/second

achieves about 10 billion comparisons per second on SOSA while using the same area as a P4 (the P4, I-SS, and 16-CMP achieve about 0.5, 2, and 32 billion comparisons per second, respectively). There are qualitatively (not quantitatively) similar results for bin-packing. SOSA's ability to exploit data parallelism in these workloads helps it outperform conventional architectures.

8.5.7 Sensitivity Analysis

This section quantifies the effect of various optimizations and changes in system parameter values on the performance of SOSA. The first is the effect of the PE length optimizations. Next, the effects of various software optimizations (sync reuse and register specifier reuse) that reduce the number of instruction bits broadcast is examined. The following section describes the effect of 1- or 2-bit-wide registers on performance. Next, the effect of different compute and communication latencies on performance is presented. Finally, the impact of various instruction buffer sizes and various node operating speeds is examined.

8.5.7.1 PE Length Optimization

Section 8.4.2 described a mechanism to limit the length of PEs in order to improve system performance. Two representative benchmarks were selected: (1) matrix multiplication for workloads that require monolithic cells; and (2) TEA for workloads that require multiple cells. Figure 8.10 plots the number of nodes required for 32 × 32 matrix multiplication (1,024 PEs) and TEA (64 PEs) as the maximum permitted PE length is varied in multiples of the ideal PE length (ideal PE length = 2 + data width / bits per register; Inf corresponds to no restriction on PE length). The results are normalized to the

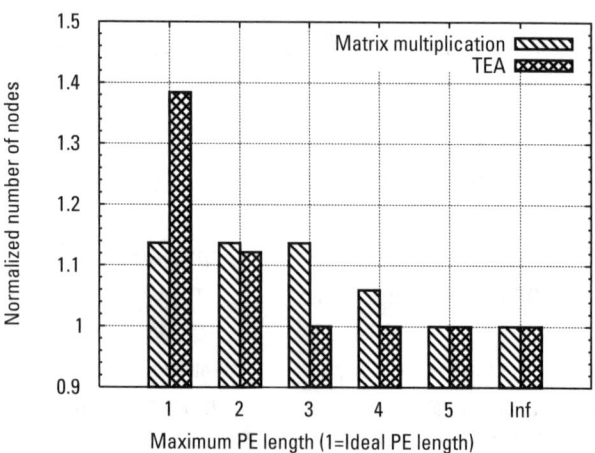

Figure 8.10 PE length versus number of nodes.

number of nodes required if there is no constraint on PE length. The results show that as PE length is restricted, the number of nodes required increases for both benchmarks (up to 14% for matrix multiplication; up to 38% for TEA). Figure 8.11 plots the running time for both benchmarks normalized to a configuration with no restrictions on PE length. As expected, limiting PE length reduces program running time (up to 14% for matrix multiply; up to 22% for TEA). However, this increased performance comes at a cost of reduced node utilization as some nodes are now unused. For workloads that use multiple cells, this also implies a reduction in the number of available cells (since each cell is larger), which is likely to reduce system throughput. Designers can strike a balance between improved performance and extra nodes required by limiting PE length, as described in Section 8.4.2.

8.5.7.2 Instruction Reuse

The results presented so far show the best performance of the SIMD architecture on matrix multiply, with instruction reuse allowed. In this section, the benefits of instruction reuse can be quantified by using matrix multiplication. Figure 8.12 plots the run-time of matrix multiply normalized to a configuration without hardware support for instruction reuse. The base configuration includes hardware to optimize the PE-Shift and uses partial broadcast of instructions. Three cases are evaluated in addition to the base case, the first with hardware support for sync reuse, the second with hardware support for register increment/decrement, and the third with both. The two bars for each configuration represent the results for 32 × 32 and 64 × 64 matrices. Both reuse optimizations reduce the bandwidth requirement of the system by re-

A Self-Organizing Defect Tolerant SIMD Architecture

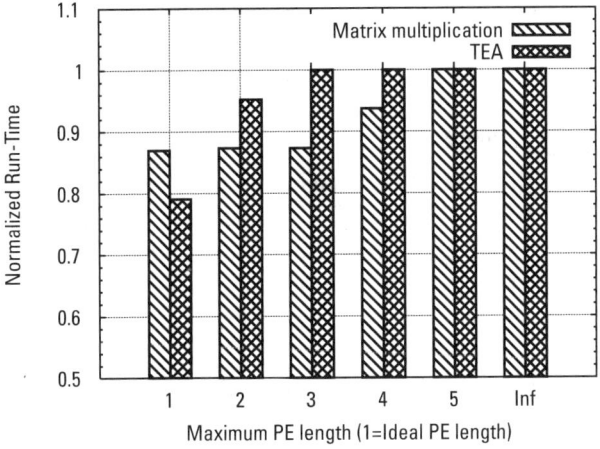

Figure 8.11 Effect of PE length optimization on program running time.

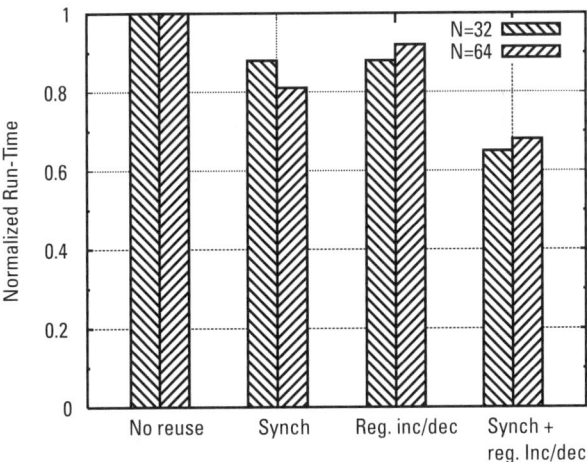

Figure 8.12 Effect of instruction reuse.

ducing the number of instruction bits broadcast. From these experiments, it is clear to see that program run-time decreases by 12% and 19% for $N = 32$ and $N = 64$, respectively, if the sync microinstruction is reused. Adding support for register increment/decrement decreases program run time by 12% for a 32×32 matrix, and by 8% for a 64×64 matrix. The larger matrix multiply is affected less because the run-time of the program is dominated by PE-Shifts, which do not benefit from the optimization. If designers enable both optimizations, run-time decreases by about 35%. A system with both optimizations presents more opportunities to reduce the number of instruction bits broadcast, and clearly benefits more than a system with any one of the optimizations.

8.5.7.3 Sensitivity to Register Width

Increasing the width of the register file increases the work done within a node per instruction. It also reduces the number of registers available to the programmer (since the total storage on the node is assumed to be fixed at 32 bits). To avoid a very small register file, only 1-bit or 2-bit wide registers are considered. Increasing the width of the register file requires time-multiplexing of a 1-bit ALU, or the use of a 2-bit-wide ALU. System performance is measured under both cases. The normalized running times for matrix multiplication and TEA are plotted in Figure 8.13. Both cases demonstrate that 2-bit-wide regis-

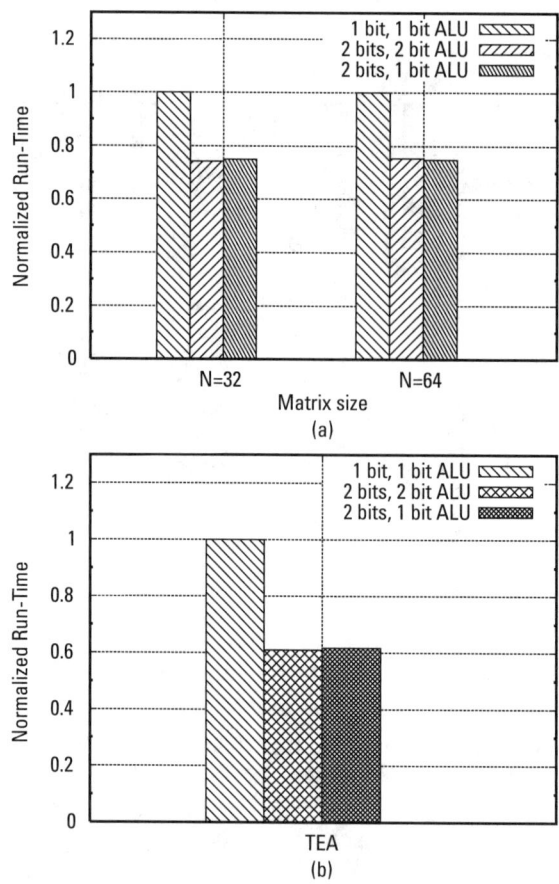

Figure 8.13 Sensitivity to register width: (a) matrix multiplication; and (b)TEA.

ters reduce program running time. In addition to reducing running time, 2-bit wide registers also reduce the number of nodes required to create a 32-bit PE by 88% (from 34 down to 18). The reduction in running time occurs for a 2-bit-wide ALU as well as for the reuse of a 1-bit-wide ALU.

8.5.7.4 Sensitivity to Compute and Communication Latencies

The effect of increasing the latency of the control/compute logic of the node can be measured. So far, the assumption has been that all activity within a node takes exactly 1 time unit. Matrix multiplication and TEA are used to evaluate the effect of increasing the latency of the control/compute logic block as well as the communication latency between the compute logic and transceivers. The normalized running time for matrix multiply and TEA for varying latencies are plotted in Figure 8.14 (a, b), respectively. For both benchmarks, observe that system performance is fairly insensitive to increased latencies less than 4 time quanta. When the total latency of the two logic blocks is greater than the latency of a bit transfer a significant drop in performance is evident as the latencies of all instructions increase.

8.5.7.5 Impact of Instruction Buffer Size

The instruction buffer stores instructions before the node is ready to execute them. It also enables the instruction broadcast mechanism to propagate instructions down the broadcast tree. Increasing the size of the instruction buffer typically improves performance since it allows increased overlap of communication and computation. However, it can cause increased contention on the bandwidth constrained links, leading to a loss in performance. In addition, increasing instruction buffer size introduces additional complexity into the node. Figure 8.15 is a plot of the normalized running time of matrix multiplication (64 × 64) and TEA as the number of entries in the instruction buffer is varied from 1 to 16 (with instruction reuse optimizations enabled). For TEA, adding instruction buffer entries improves performance, but results in diminishing gains beyond four instruction buffer entries. Matrix multiplication experiences an increase in running time beyond one entry due to increased network contention. A single entry instruction buffer is used as a trade-off between node complexity and performance improvement over a node design without the instruction buffer.

8.5.7.6 Effect of Increasing Operating Speed

The results presented in the previous section assumed a conservative value of 1 ns for the time unit. Recent measurements of carbon nanotubes indicate that it may be possible to operate devices based on nanotubes at very high frequencies (~1 terahertz) [20, 21]. Figure 8.16 shows the run time for the

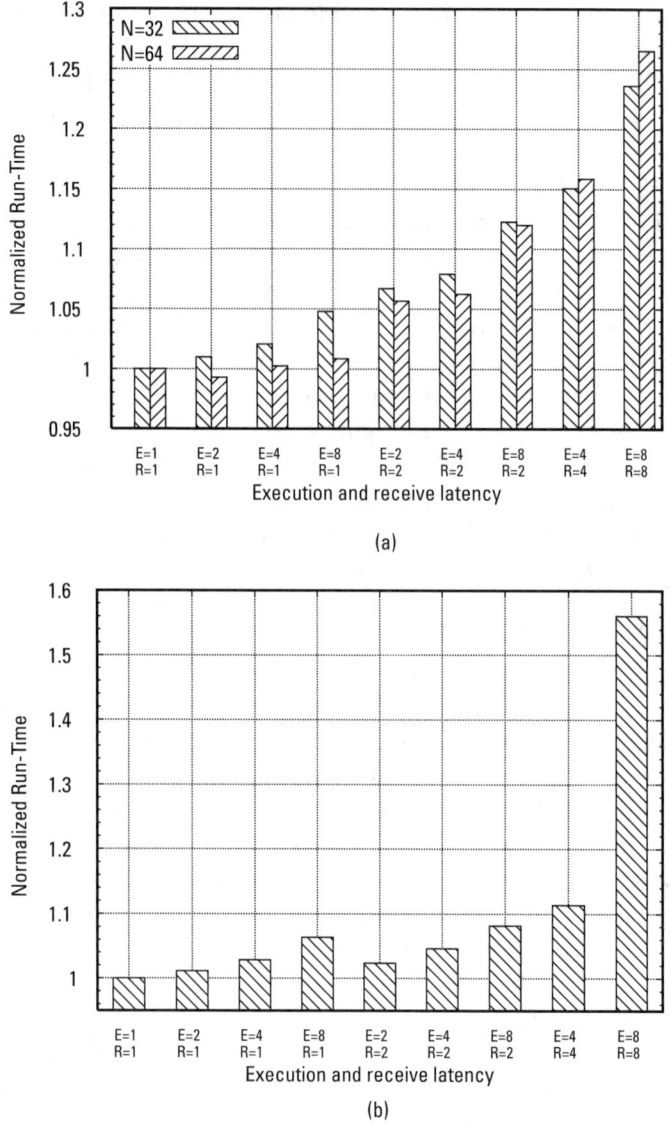

Figure 8.14 Sensitivity to execution and communication latencies: (a) matrix multiplication; and (b) TEA.

matrix multiply for two matrix sizes ($N = 128$, $N = 512$) for different time unit values. The running time for the Pentium 4 running at 3 GHz is also shown as a point of comparison. The figure shows that if SOSA could operate with lower values for the time unit, it would achieve run-times closer to the Pentium 4 for smaller matrix sizes ($N = 128$, with a time unit of ~100 ps).

A Self-Organizing Defect Tolerant SIMD Architecture

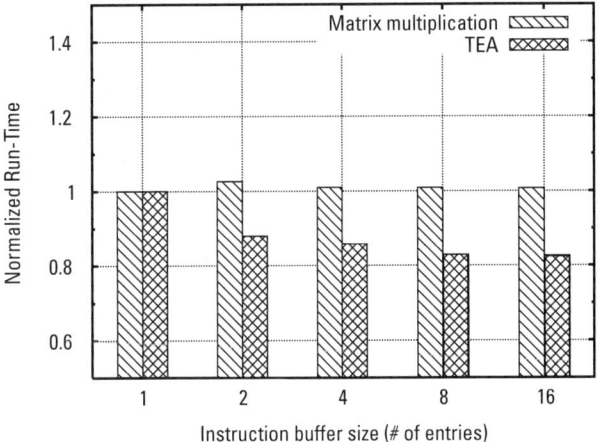

Figure 8.15 Effect of instruction buffer size.

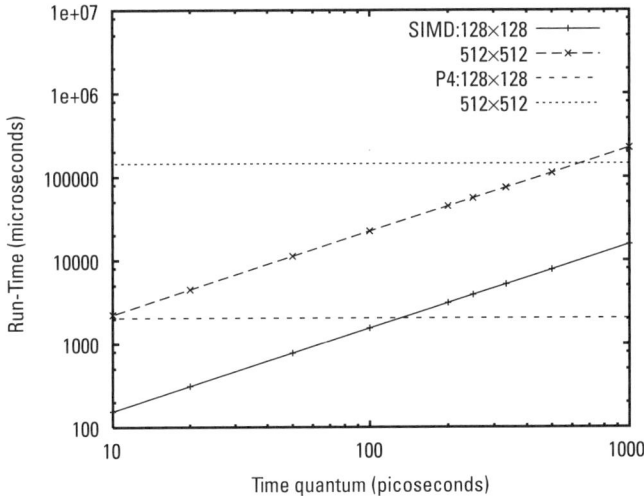

Figure 8.16 Effect of operating speed.

8.5.7.7 Sensitivity Summary

Sensitivity analysis shows that SOSA's performance is not very sensitive to compute and internal communication latencies as long as these latencies are greater than internode communication latencies. Increasing the size of the instruction buffer can improve performance, but this results in increased node complexity. SOSA's performance improves if wider registers are used, which also leads to a reduction in the number of nodes required to form a PE. However, due to node size limitations, there is a trade-off between wider registers and number of registers available. SOSA can also benefit from running at faster speeds, limiting PE lengths and the instruction reuse mechanisms. Next, a critical aspect of SOSA's design is examined: its ability to tolerate defective nodes.

8.5.8 Defect Tolerance

The ability to tolerate defects is one of the primary features of SOSA. To test the defect tolerance and to measure the effect of defects on performance, the node defect rate is varied. First, an examination of the effect of defects on the throughput of a system configured into multiple cells is presented. If the total system area is constant (100 mm^2), as node defect rates increase fewer cells can be configured, resulting in reduced throughput. Figure 8.17 plots the throughput for TEA and XTEA, as node defect rates increase from 0% to 30% revealing a graceful degradation in performance. The connectivity of the random network of nodes is severely affected by node defect rates greater than 30%. This results in network partitions with insufficient functioning nodes in each partition to configure a 64-PE cell.

For single cell applications, the entire system must be overprovisioned to ensure that a sufficient number of PEs can be configured. Thus, defects indirectly impact performance by reducing network connectivity and band-

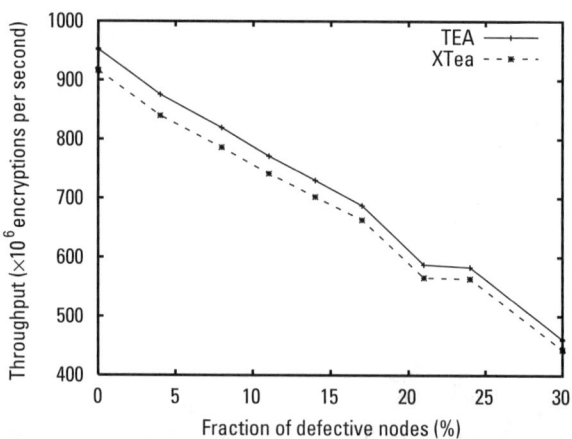

Figure 8.17 TEA/XTEA—graceful degradation of throughput with increasing node defect rate.

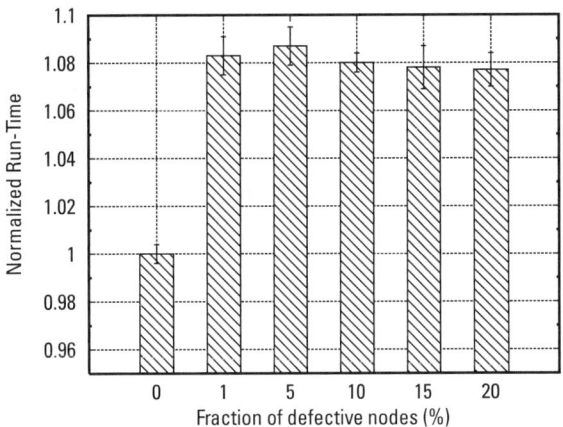

Figure 8.18 Matrix multiplication—effect of defects on run-time.

width. In all experiments, SOSA has 30% more nodes (24,000 total nodes) than the minimum needed for a 32 × 32 matrix multiply. Figure 8.18 shows the running time for 32 × 32 matrix multiplication as the number of defective nodes is increased from 0% to 20%. The results indicate that the running time increases by about 8% (compared to a case with no defects), primarily because the average length of PEs increases. Results for the other workloads are qualitatively similar. If the system cannot configure sufficient PEs, the problem could potentially be divided into parts that can be solved with the available PEs. Though the defect tolerance capabilities of the RPF algorithm have been demonstrated before, these experiments show that the ability to tolerate high defect rates incurs only a small performance penalty (~8% for N = 32, 32-bit PEs), a characteristic of increasing importance for future systems.

8.5.9 Result Summary

The results in this section show that a system built using a random network of simple nodes can outperform a P4 and an I-SS processor, despite being severely bandwidth limited and operating devices at a lower switching speed. A scaled-up version of the system can outperform an ideal 16-way CMP. The results also highlight SOSA's flexibility in configuring independent cells to improve system utilization and throughput. SOSA provides higher throughput than the P4 and I-SS while using the same area. Coupled with the ability to tolerate a significant defect rate, SOSA shows potential in harnessing the higher device densities that emerging technologies promise to deliver.

8.6 Limitations and Future Work

The performance evaluation reinforces the common knowledge that a high computation-to-communication ratio is critical for achieving good performance, particularly on SOSA due to its low bandwidth and high communication latencies. SOSA is likely to achieve good performance on pipelined implementations of programs that require high throughput, or programs that require little inter-PE communication, nearest neighbor communication, or regular and unidirectional dataflow. In contrast, SOSA is unlikely to achieve good performance for programs that require all-to-all communication because of the logical ring topology and limited network bandwidth. Although SOSA achieves good performance on most of the workloads studied, it is not a general purpose architecture (as clearly demonstrated by the performance of sort). SOSA is unlikely to be able to match the performance of conventional processors on most general purpose workloads. SOSA is also limited by lack of hardware support for floating point operations. There are software implementations of floating point operations, but performance is limited by the use of predicated instructions to handle control dependencies between different parts of the operations.

There are a number of avenues to extend the work on SOSA: extending SOSA to speed up floating point operations, exploit multiple anchors to increase system I/O bandwidth, and to handle transient faults through redundant execution or by extending PEs to perform simple checksum/parity computations. Another direction is to extend the software tool-chain to explore compiler optimizations. Other open research areas include modifications to the configuration mechanism to exploit unused links to improve I/O bandwidth, configuring nodes for specific functionality (e.g., floating point or storage), using SOSA as an add-on to a conventional core to improve performance on data parallel workloads, and creating hybrid cores that mix CMOS and self-assembled devices.

As self-assembly technology matures, some of the severe fabrication limitations may be removed. The performance of I-SOSA provides an upper bound of SOSA performance, assuming a time quantum of 1 ns. However, with fewer fabrication limitations, it might be possible to achieve better performance by revisiting design decisions that trade off performance for reduced design complexity. For example, if fabrication enables larger nodes, it might be possible to fit a full PE in one node, or to build more complex transceivers to achieve better network connectivity [28]. As emerging device technologies improve, it may be possible to operate them at higher speeds (causing a potential increase in power consumption). It is important to note that while SOSA is motivated by DNA-based self-assembly as the fabrication process,

SOSA is applicable to any manufacturing technique that results in high defect rates and a loss of precise control during parts of the fabrication process.

8.7 Related Work

There is a large body of research on building computing systems with similar goals, but this research differs primarily in the granularity of the basic computational blocks used to form the system. SOSA must use very simple computational nodes due to fabrication constraints. This section focuses on closely related work applicable to emerging technologies. The decoupled array multiprocessor [29] (Chapter 6) and the nanoscale active network architecture [15] (Chapter 7) use DNA-based self-assembly of nanoelectronic devices. The DAMP exploits data parallelism, but it is not capable of efficient data exchange between processing elements, limiting it to embarrassingly parallel problems. SOSA uses more sophisticated self-organization and achieves better performance than NANA since it has lower communication overheads and better node utilization, and uses a single node type.

Researchers have developed FPGA-based reconfigurable architectures [30–32] that extract a system-level defect map and use this external map to configure the system, while isolating defective regions. The key difference is that SOSA configures higher level logic blocks (nodes as opposed to gates in an FPGA) and does not require an external defect map. This is critical since there is little information available to designers about the physical network topology. Researchers have proposed various voting and redundancy schemes to deal with defects, including triple modular redundancy (TMR) [33], N modular redundancy [34], NAND multiplexing and hot/cold sparing [35] (particularly in the context of molecular electronic systems). The defect tolerance scheme used in this work does not rely on redundant computation but isolates defective regions in the system. There is extensive research on designing and building vector [36, 37] and SIMD machines [38, 39], including the cell processor [40]. The cell processor has eight SIMD cores that can be programmed independently, unlike the PEs in SOSA. The primary difference between SOSA and past work is the focus on overcoming the challenges imposed by the fabrication technology and the need to tolerate defects.

8.8 Conclusions

With the expected rise in defect rates as device sizes shrink, defect tolerance will be a critical requirement for future system architectures. These increasing defect rates will contribute directly to the exponentially increasing cost of top-down manufacturing. The use of bottom-up techniques like self-assembly will help lower costs but may also result in higher defect rates and a loss of precise control over the manufacturing process. This makes it imperative for architects to develop defect tolerant architectures to exploit the full potential of future nanoscale devices. This chapter presents SOSA, a self-organizing SIMD architecture built from a random network of simple computational nodes. Despite high defect rates, low bandwidth, and lack of underlying physical structure, for data parallel workloads, SOSA is able to perform better than conventional superscalar processors, while operating at a lower speed and consuming much less power. A scaled version of SOSA can perform better than an ideal 16-way CMP. As the underlying technology matures, SOSA's performance can be further improved as fabrication limitations are removed. While SOSA does not solve all problems encountered with self-assembled architectures, it is a step towards realizing defect-tolerant computing systems built using emerging technologies that may provide inexpensive terascale integration.

8.9 Programming SOSA—Matrix Multiplication

This section provides a brief overview of programming SOSA using matrix multiplication as a running example, and demonstrates how various optimizations can be applied to improve performance. The N^3 algorithm for multiplying two $N \times N$ matrices A and B is shown in Figure 8.19.

Since SOSA does not include memory addressable from within the PEs, assume that data is distributed among the PEs with a simple data layout— each PE holds one element each of the input matrices (depicted in Figure 8.20, for two 4×4 matrices). Next, divide the algorithm into four parts, each of which is repeated N times. The first part computes the N^3 products; the second part accumulates sums to create elements of the result; the third part moves data within the PEs to set up the next iteration; and the fourth part moves each newly computed element of the result to its final location. Since SOSA does not have a native multiplication instruction, the first part is not trivial, and it is implemented using a shift-add algorithm.

```
for i=j to N
  for j=1 to N
    for k=1 to N
      C[i][j]=C[i][j]+A[i][j]*B[i][k];
    End
  End
End
```

Figure 8.19 Matrix multiplication pseudo-code.

Figure 8.21 shows the first version of the primary matrix multiply loop. There are four components as stated earlier: multiply, accumulate, align data, move result. The largest fraction of running time is spent in the first two parts of the algorithm, and so this is the focus of optimization efforts. The primary optimizations applied to the third and fourth part include the reuse of microinstructions where possible.

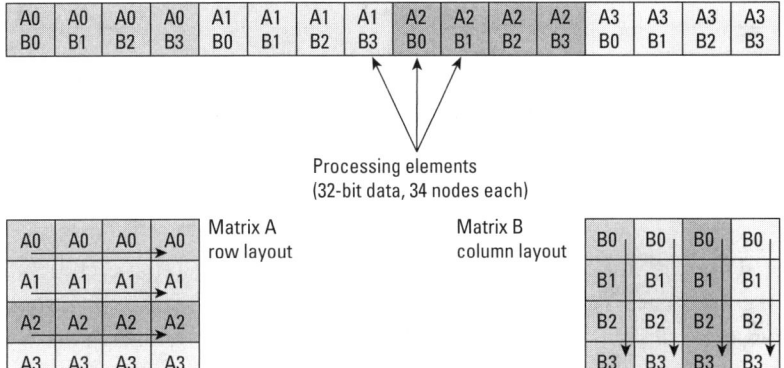

Figure 8.20 Matrix layout.

```
; Initialize before Multiply
CPREG R4,R2   ; Copy R4 -> R2
CPREG R3,R1   ; Copy R3 -> R1
CLEAR R5      ; Clear R5
; Multiply (Loop Dw times) (Dw: Data Width)
SHIFTLM R1    ; Shift LSB to MSB (multiply by 2)
PSHIFTML R2,R5 ; Shift MSB to LSB, LSB to pred.reg R5
PRADD R5,R1,R5 ; if predicate is set, R5=R5+R1
CLEAR R6      ; Clear R6
; Accumulate partial products
;---Repeat N times---
ADD R6,R6,R5  ; Accumulate partial sum
CPREG R6,R5   ; Copy R6 to R5
SHIFTMLPE R5  ; Send accumulated sum to previous PE
; Align rows of matrix A for next set of multiplies
;(Repeat (Dw+2)*N times)
SHIFTMLPE R4  ; Move A 'N' PEs to the left
; Move Result
CPREG R8,R9   ; if R8==1, this PE holds the first
              ; element of a row/column, move this to R9
PSHIFTML R9,R6 ; Move that bit into the predicate register R6
PRCPREG R6,R7 ; if predicate set, copy R6 -> R7
SHIFTMLPE R7  ; Move R7 one PE to the left (*(Dw+2))
```

Figure 8.21 Matrix multiply assembly code—no optimizations.

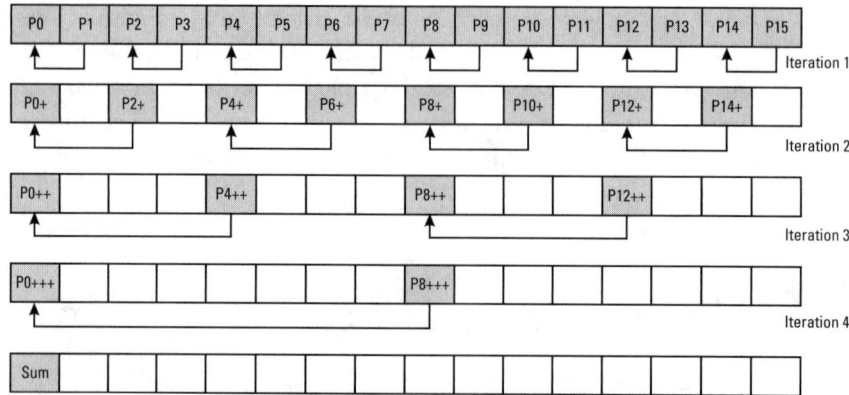

Figure 8.22 Logarithmic accumulate.

The accumulate operation is optimized by observing that each iteration must accumulate N products into a single sum. However, matrix sizes that are a power of 2 enable optimizing this accumulation step by replacing the N add iterations by $\log(N)$ iterations, and in every kth iteration, moving the sum 2^k PEs before performing the accumulate. This is depicted in Figure 8.22 for $N = 16$. This reduces the number of iterations but does not reduce the amount of data that must be communicated. Note some extra ADD instructions are performed on data elements that do not contribute to the final result. The final accumulate code is shown in Figure 8.23.

Loop unrolling is used to optimize the multiplication and maximize the use of the register file within each node. With 1-bit-wide registers, the multiply loop can be unrolled 16 times and can perform only two iterations of shift-add. Each unrolled iteration creates a shifted version of the multiplicand, and generates predicate bits using the multiplier. A predicated add controls whether the shifted multiplicand gets added depending on the predicate bit created by the multiplier. The loop unrolling allows reusing microinstructions, which helps reduce instruction execution time.

```
; Accumulate partial products
;---Repeat log2(N) times---
ADD R6,R6,R5    ; Accumulate partial sum
CPREG R6,R5 ;   Copy R6 to R5
SHIFTMLPE R5    ;   For iteration i, repeat (Dw+2)*i*2 times
; End Repeat
```

Figure 8.23 Logarithmic accumulate—assembly code.

References

[1] Robinson, B. H., and N. C. Seeman, "The Design of a Biochip: A Self-Assembling Molecular-Scale Memory Device," *Protein Engineering*, Vol. 4, No. 1, 1987, pp. 295–300.

[2] Patwardhan, J. P., et al., "Circuit and System Architecture for DNA-Guided Self-Assembly of Nanoelectronics," *Proc. Foundations of Nanoscience: Self-Assembled Architectures and Devices*, 2004, pp. 344–358.

[3] Winfree, E., et al., "Design and Self-Assembly of Two-Dimensional DNA Crystals," *Nature*, Vol. 394, 1998, pp. 539–544.

[4] Yan, H., et al., "DNA Templated Self-Assembly of Protein Arrays and Highly Conductive Nanowires," *Science*, Vol. 301, 2003, pp. 1882–1884.

[5] Bachtold, A., et al., "Logic Circuits with Carbon Nanotube Transistors," *Science*, Vol. 294, 2001, pp. 1317–1320.

[6] Dwyer, C., et al., "DNA Functionalized Single-Walled Carbon Nanotubes," *Nanotechnology*, Vol. 13, 2002, pp. 601–604.

[7] Huang, Y., et al., "Logic Gates and Computation from Assembled Nanowire Building Blocks," *Science*, Vol. 294, 2001, pp. 1313–1317.

[8] Park, S. H., et al., "Finite-Size, Fully-Addressable DNA Tile Lattices Formed by Hierarchical Assembly Procedures," *Angewandte Chemie*, Vol. 45, 2006, pp. 735–739.

[9] Dwyer, C., et al., "The Design and Fabrication of a Fully Addressable 8-tile DNA Lattice," *Proc. Foundations of Nanoscience: Self-Assembled Architectures and Devices*, 2005, pp. 187–191.

[10] Skinner, K., et al., "Nanowire Transistors, Gate Electrodes, and Their Directed Self-Assembly," *Proc. 72nd Southeastern Section of the American Physical Society (SESAPS)*, 2005, p. H2+.

[11] Abelson, H., et al., "Amorphous Computing," *Communications of the ACM*, Vol. 43, 2000, pp. 74–82.

[12] Schroeder, M. D., et al., "Autonet: A High-Speed, Self-Configuring Local Area Network Using Point to Point Links," *IEEE Journal on Selected Areas in Communications*, Vol. 9, 1991.

[13] Intanagonwiwat, C., R. Govindan, and D. Estrin, "Directed Diffusion: A Scalable and Robust Communication Paradigm for Sensor Networks," *6th Annual International Conference on Mobile Computing and Networking*, 2000, pp. 56–67.

[14] Mai, K., et al., "Smart Memories: A Modular Reconfigurable Architecture," *Proc. 27th Annual International Symposium on Computer Architecture*, 2000, pp. 161–171.

[15] Patwardhan, J. P., et al., "NANA: a Nano-Scale Active Network Architecture," *J. Emerg. Technol. Comput. Syst.*, Vol. 2, No. 1, 2006, pp. 1–30.

[16] Dwyer, C., M. Cheung, and D. J. Sorin, "Semi-Empirical SPICE Models for Carbon Nanotube FET Logic," *Proc. IEEE Conference on Nanotechnology*, 2004, pp. 386–388.

[17] Dalal, Y. K. and R. M. Metcalfe, "Reverse Path Forwarding of Broadcast Packets," *Communications of the ACM*, Vol. 21, No. 12, 1978, pp. 1040–1048.

[18] Patwardhan, J. P., et al., "Evaluating the Connectivity of Self-Assembled Networks of Nanoscale Processing Elements," *Proc. IEEE International Workshop on Design and Test of Defect-Tolerant Nanoscale Architectures (NANOARCH '05)*, 2005, pp. 2.1–2.8.

[19] Patwardhan, J. P., C. Dwyer, and A. R. Lebeck, "Design and Evaluation of Fail-Stop Self-Assembled Nanoscale Processing Elements," *Proc. IEEE International Workshop on Design and Test of Defect-Tolerant Nanoscale Architectures (NANOARCH '06)*, 2006.

[20] Burke, P. J., "Carbon Nanotube Devices for GHz to THz Applications," *SPIE*, Vol. 5593, 2004, pp. 52–61.

[21] Rosenblatt, S., et al., "Mixing at 50GHz Using a Single-Walled Carbon Nanotube Transistor," *Applied Physics Letters*, Vol. 87, 2005, pp. 153111.

[22] Lines, A., "Asynchronous Interconnect for Synchronous SoC Design," *IEEE Micro*, Vol. 24, No. 1, 2004, pp. 32–41.

[23] Austin, T., E. Larson, and D. Ernst, "SimpleScalar: An Infrastructure for Computer System Modeling," *IEEE Computer*, Vol. 35, No. 2, 2002, pp. 59–67.

[24] Performance Database Server. http://www.netlib.org/performance/html/PDStop.html

[25] Knuth, D. E., *The Art of Computer Programming*, Reading, MA: Addison-Wesley, 1973.

[26] Wheeler, D., and R. Needham, "TEA: A Tiny Encryption Algorithm," *Proc. Fast Software Encryption: Second International Workshop*, 1994.

[27] Needham, R. and D. Wheeler, "TEA Extensions," *Technical Report, University of Cambridge*, 1997.

[28] Patwardhan, J. P., C. Dwyer, and A. R. Lebeck, "Self-Assembled Networks: Control vs. Complexity," *Proc. First International Conference on Nano-Networks (NANONETS)*, 2006.

[29] Dwyer, C., et al., "DNA Self-assembled Parallel Computer Architectures," *Nanotechnology*, Vol. 15, 2004, pp. 1688–1694.

[30] Goldstein, S. C., and M. Budiu, "NanoFabrics: Spatial Computing Using Molecular Electronics," *Proc. 28th Annual International Symposium on Computer Architecture (ISCA)*, 2001, pp. 178–191.

[31] Heath, J. R., et al., "A Defect-Tolerant Computer Architecture: Opportunities for Nanotechnology," *Science*, Vol. 280, 1998, pp. 1716–1721.

[32] Culbertson, W. B., et al., "The Teramac Custom Computer: Extending the Limits with Defect Tolerance," *Proc. of the IEEE Int'l Symposium on Defect and Fault Tolerance in VLSI Systems*, 1996.

[33] Lyons, R. E., and W. Vanderkulk, "The Use of Triple-modular Redundancy to Improve Computer Reliability," *IBM Journal*, 1962, pp. 200–209.

[34] von Neumann, J., C. E. Shannon, and J. McCarthy, "Probabilistic Logics and the Synthesis of Reliable Organisms from Unreliable Components," *Automata Studies*, 1956, pp. 43–98.

[35] DeHon, A., "Array-Based Architecture for Molecular Electronics," *Proc. of the First Workshop on Non-Silicon Computation (NSC-1)*, 2002.

[36] Ciricescu, S., et al., "The Reconfigurable Streaming Vector Processor (RSVP)," *Proc. 36th Annual IEEE/ACM International Symposium on Microarchitecture*, 2003, pp. 141–150.

[37] Espasa, R., et al., "Tarantula: A Vector Extension to the Alpha Architecture," *Proc. 29th Annual International Symposium on Computer Architecture*, 2002, pp. 281–292.

[38] Tucker, L., and G. Robertson, "Architecture and Applications of the Connection Machine," *IEEE Computer*, Vol. 21, 1988, pp. 26–38.

[39] Ujval, K., et al., "The Imagine Stream Processor," *Proc. IEEE International Conference on Computer Design*, 2002, pp. 282–288.

[40] Hofstee, H. P., "Power Efficient Processor Architecture and the Cell Processor," *Proc. 11th International Symposium on High-Performance Computer Architecture (HPCA)*, 2005, pp. 258–262.

9

Overcoming Randomness with Increased Complexity

9.1 Introduction

DNA-guided self-assembly of nanoelectronic components is a promising technology that could usher in an era of tera- to petascale integration and extend Moore's law beyond the capabilities of CMOS. A key advantage of this technology is its ability to manufacture a large number of circuit blocks in parallel. The previously proposed circuit architecture (see Chapter 3) could be used to manufacture computing circuits using this technology. However, systems designed for this technology must address the increased randomness inherent to self-assembly.

The oracle and DAMP systems exploit randomness to generate values that statistically cover a large space. In contrast, NANA and SOSA do not rely on randomness, but instead incorporate tolerance to randomness into their designs. For example, they incorporate built-in self-test to provide fail-stop node behavior in the presence of defects (e.g., a node that is not self-assembling correctly) and they tolerate arbitrary physical network topologies generated by self-assembled links between nodes.

This chapter takes a closer look at two aspects related to randomness: (1) the design of fail-stop SOSA nodes, and (2) the trade-off between node complexity and control over node positioning and connectivity during fabrication.

We begin by exploring the trade-offs in implementing test mechanisms to achieve fail-stop behavior in nodes while meeting manufacturing constraints. The design uses hardware self-test mechanisms to verify critical node components, as well as software tests for noncritical components. The design also reuses test logic where possible and moves noncritical verification to software in order to meet technological size constraints. The modularity of the node and test logic, and the ability to disable defective components enables the use of nodes with some (noncritical) defective components. This allows the system to tolerate higher transistor defect rates. In particular, if nodes with at least one communication unit and one compute unit, or two communication units, are allowed to operate, the system can tolerate a transistor defect probability of 1.5×10^{-4}. This is an order of magnitude higher than the defect probability that can be tolerated when a single defective transistor results in an unusable node.

The second part of this chapter explores the impact of randomness during fabrication, specifically node networks created by varying control over three aspects of the self-assembly process (node placement, node orientation, and internode link creation). In particular, this work examines the trade-off between node complexity and control required during self-assembly to maximize the number of connected nodes in the network. As the level of control decreases, node communication hardware needs to be augmented to allow link sharing between several transceivers. This also results in better network connectivity in the presence of defective nodes and links. Finally, with enough available nodes, the specific network topology has a negligible effect on SOSA performance.

9.2 Design and Evaluation of Fail-Stop Nodes

A system architecture that uses DNA self-assembly must explicitly incorporate defect tolerance strategies. Chapter 7 presented the reverse path forwarding mechanism to isolate defective nodes in a random network of self-assembled nodes. However, this assumes fail-stop nodes that are defective if there is a single transistor defect, which leads to unusable systems if the per-device defect probability is greater than 4×10^{-5}.

This chapter explores strategies for implementing fail-stop processing elements (nodes) within the constraints imposed by self-assembly. The design extends the defect isolation mechanism to operate within a node and uses a combination of hardware and software test strategies to verify the operation of node components. If a node component fails or never completes the test, it is assumed to be defective and is not used, resulting in fail-stop behavior. Distinct tests for different node components enable the use of nodes with some defective components, as long as the defects do not affect critical functionality. Partially functional nodes can help the system tolerate a higher transistor defect probability (1.5×10^{-4}) and improve system connectivity as node defect rates increase. The primary contributions of this work are:

- Implementing fail-stop nodes using a combination of hardware and software test strategies to verify the operation of node components;

- Extending a previously proposed defect isolation mechanism to improve system connectivity and increase tolerance of higher device defect probabilities.

The rest of this section is organized as follows. A brief description of the system and defect isolation algorithm is presented in Section 9.2.1. The node architecture and test strategies are discussed in Section 9.2.2 and evaluated in Section 9.2.3. Section 9.2.4 discusses related work, and the section concludes in Section 9.2.5.

9.2.1 Defect Isolation Using Reverse Path Forwarding

This section provides a brief summary of a target system with the defect isolation mechanism. One of the key advantages of self-assembly is its ability to construct a large number of devices in parallel. The circuit and system architectures described in the previous chapters, specifically SOSA (Chapter 8), use DNA-guided self-assembly to build a patterned scaffold with selective placement of carbon nanotube transistors or nanorods as active devices. This enables the construction of a large number of small circuit blocks (nodes) that can be connected using metallized DNA to form a random network. Expected limitations of self-assembly constrain node size, and the design assumes nodes with approximately 10,000 usable transistors. Although DNA-based self-assembly provides a great degree of control during design through base-pair specification, large-scale self-assembly introduces randomness (i.e., defects) both within nodes (e.g., inexact base pairing) and in the interconnection of nodes (e.g., early termination of link growth).

An adaptation of the reverse path forwarding (RPF) algorithm [1] isolates defective nodes in the network and organizes functional nodes [2]. A brief overview of the defect isolation mechanism is provided here; a more detailed discussion is found in Chapter 7 and elsewhere [2]. The algorithm begins with a single "gradient" packet inserted from the external microscale through an anchor. A node receiving this packet for the first time performs two actions: (1) it notes the input link on which the packet arrived (i.e., the gradient), and (2) it forwards the packet on its active links, except the input link. If a node receives the packet again, it simply discards it. This results in a rapid broadcast of the packet to all nodes in the system. A node that receives the packet has a known route to the anchor where the packet was inserted by following the gradient through intermediate nodes. At the end of the algorithm, all functional nodes that received the gradient are connected on a tree (broadcast tree). The design makes a key assumption in this process: nodes that propagate the broadcast are defect free (or, defective nodes are fail-stop and do not participate in the broadcast). This results in the isolation of defective nodes (which do not propagate the broadcast) since no other functional node has a route to the anchor through a defective node. If defective nodes propagate the gradient broadcast, the system could be misconfigured and not function correctly. Thus, each node must implement fail-stop behavior.

9.2.2 Fail-Stop Nodes

This section explores hardware and software test strategies that can help achieve fail-stop behavior. A node can be divided into three main components: (1) communication logic, (2) configuration logic, and (3) compute logic; each component can have an independent test strategy. This simplifies test logic and enables the use of a partially functional node by isolating components that do not pass logic tests. Previous work made the conservative assumption that a node with a single defect is unusable. The ability to use partially functional nodes allows different node failure modes that can better utilize the defect-free parts of a node.

This section begins with a description of node architecture (Section 9.2.2.1) and identifies logic blocks that are critical to achieving fail-stop behavior (Section 9.2.2.2). It examines different hardware/software design options for implementing fail-stop, and identifies the benefits of each approach (Section 9.2.2.3). Next, is a description of the test mechanisms for communication (Section 9.2.2.4), configuration (Section 9.2.2.5) and compute logic (Section 9.2.2.6). The effect of using such partially defective nodes on device reliability requirements, defect isolation, and system operation is then explored (Section 9.2.2.7).

9.2.2.1 Node Architecture

To review, each SOSA node is an asynchronous circuit and can be divided into three parts: communication logic, configuration logic, and compute logic. The communication logic consists of four transceivers that allow the node to communicate with up to four neighbors on single bit links. Transceivers use a four-phase handshake protocol for data transfer over links. Each handshake transfers 1 bit, and links support full-duplex data transfer. Each transceiver supports three virtual channels [3] using 1-bit buffers. The four transceivers are connected to each other and the compute/configuration logic through point-to-point links for each virtual channel. The configuration logic is responsible for setting up internal node routing during the gradient broadcast phase. It configures virtual channel 0 (VC0) for broadcast routing, virtual channel 1 (VC1) for a depth-first traversal of the gradient broadcast tree, and virtual channel 2 (VC2) with the reverse routing order of VC1. The compu-

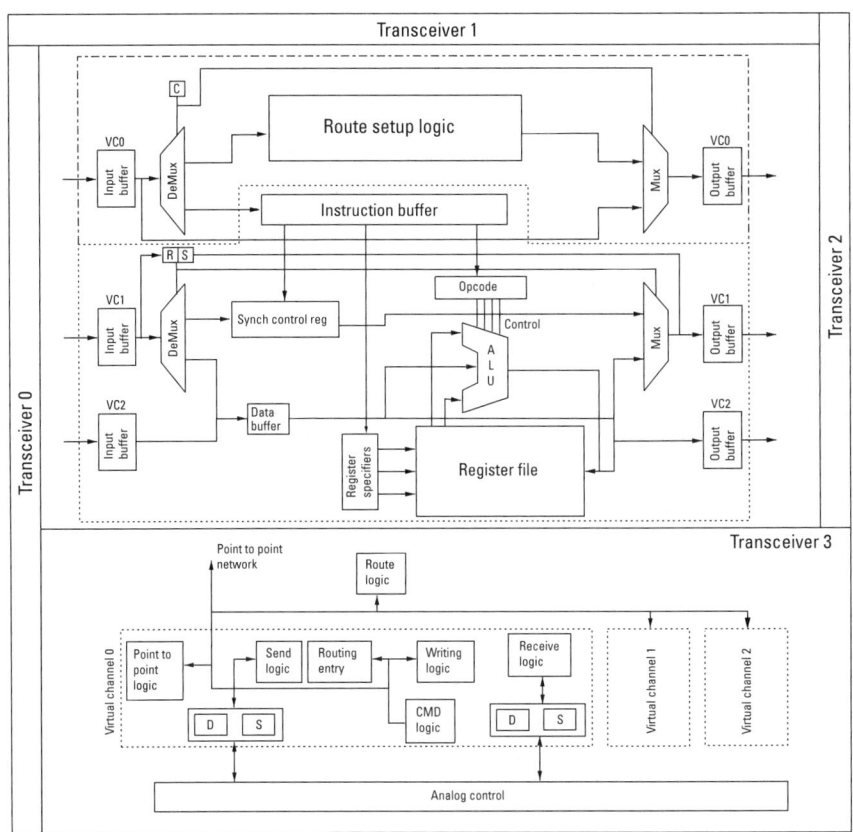

Figure 9.1 SOSA node.

tation logic implements a simple two-stage pipeline that allows the execution of instructions on 2-bit data slices. The computation logic has 32 bits of storage configured as a 16-entry 2-bit-wide register file and an ALU that can perform simple arithmetic and logic functions. Figure 9.1 shows the block diagram of a node, clearly identifying the communication, configuration, and compute logic (the figure shows details of only one transceiver). A significant fraction of the node logic is devoted to communication support (both internode and intra-node). The next step is to examine each logic block to determine its criticality with respect to achieving fail-stop behavior.

9.2.2.2 Critical Node Logic

A logic block that must be defect-free for the node to function correctly is called "critical." These logic blocks must be tested before a node accepts any external input to avoid the possibility of system misconfiguration. Logic for

Table 9.1
Node Component Classification

Component	Critical	Description
Configuration logic	Yes	Input arbitration on VC0, depth-first route setup
Transceiver logic—VC0	Yes	Send/receive logic for VC0
Transceiver logic—VC1/2	No	Send/receive logic for VC1/2
Point-to-point interconnect	VC0: Yes; VC1/2: No	Data interconnect within node
ALU	No	Arithmetic logic unit
Register file	No	Register file in compute block
Instruction buffer	No	First pipeline stage
Execution control registers	No	Storage for microinstructions

VC0 (communication logic) and route setup (configuration logic) is critical. All other logic in the node can be tested during the defect isolation phase since it does not affect the ability of a node to receive and send data. While this remaining logic is not critical, it must still be tested to ensure correctness. This can be performed with hardware or in software during defect isolation.

Table 9.1 classifies various node logic blocks based on their criticality. The classification of logic blocks into critical/noncritical provides a simple way of determining what logic should be tested in hardware and what can be tested with software. This trade-off is explored next.

9.2.2.3 Fail-Stop Node Design Options

The goal is to achieve fail-stop behavior in nodes with minimal extra hardware. Critical logic must be tested before a node communicates with its neighbors, which implies the need for hardware test logic. For noncritical logic, the designer can choose between three options: (1) hardware only test, (2) software only test, and (3) hardware-software hybrid test.

Hardware Test

The designer can add logic to each node to test the functionality of all components. This is equivalent to built-in self-test (BIST) [4, 5] that does not require external test vectors. The primary advantages of hardware testing are low latency and the ability to test the node independent of the rest of the system. However, the size of a node that relies only on hardware test circuitry would exceed technological size constraints. This makes a hardware test strategy impractical. Note that critical logic still requires hardware testing.

Software (External) Test

For all noncritical logic, the designer could rely on software-based testing using external test vectors. This can be combined with gradient broadcast to allow parallel testing of nodes, which would reduce test latency. This approach works well for instruction execution logic, but is not as useful for other components. For example, software testing of the transceiver circuitry for VC1 requires hardware support to allow routing of test vectors to the transceiver logic. For small logic blocks, this extra hardware could be more expensive than implementing a hardware test scheme.

Hardware/Software Hybrid Test

The final option for testing is to use a hybrid approach of hardware testing for simple components and software testing for more complex components. For example, transceiver logic is simple and requires identical testing for all three virtual channels. This can be done efficiently with simple test hardware. Furthermore, this test hardware can be shared between the three virtual channels. While this could increase test latency by a small amount, it results in reduced circuit size. Compute logic is fairly complex, and requires a large number of test vectors to ensure correct functionality. The designer can exploit existing hardware to test compute logic using external test vectors, with minimal extra hardware. This allows the designer to keep node size within technological constraints.

In summary, the design uses hardware test strategies for node components that can be tested with simple logic. If possible, the design reuses test circuits to minimize overhead. The next three subsections describe test strategies for the three main components in a node.

9.2.2.4 Fail-Stop Communication Logic

Communication logic within a node supports three virtual channels and has two primary components: (1) four transceivers, and (2) point-to-point links. The circuits for VC0 are part of the node's critical logic since they are required during configuration. VC1 and VC2 are not part of critical logic, but can share test logic with VC0.

Each transceiver in a node must be tested to ensure correct functionality as defective logic in a transceiver can lead to incorrect system behavior. A node can be a useful part of a larger system even if it has only one functioning transceiver. However, if there are defective transceivers in a node, it is critical to isolate them from the rest of the system. To achieve this, each transceiver is augmented with simple test logic and a loopback path between the output and input logic of each transceiver. This path is enabled during test only. The design exploits the simple four-phase handshake protocol used by the asynchronous logic in designing a test circuit that verifies the operation of the input/output logic. The transceiver is assumed to be defective by default. If the test verifies transceiver operation, the test circuit generates a signal to indicate that the transceiver is operational.

The largest component of the test logic is a 2-bit state machine which inserts a test bit pattern into the transceiver output logic. The test pattern consists of 2 bits (0 followed by 1). The test logic inserts the 0, then waits until it loops back to the input logic. If the test logic successfully receives the 0 from the input logic, it inserts a 1 and waits for it to loop back. If both data bits (0 and 1) are received correctly, the test logic generates a TEST_OK signal. If the data is never received or incorrect data is received, this signal is not generated, isolating this transceiver from the rest of the node.

In addition to testing the transceiver logic, the circuit must test the point-to-point links that connect transceivers. However, routing on the point-to-point links depends on the result of the configuration process, so the point-to-point links are tested at the same time as the configuration logic.

9.2.2.5 Fail-Stop Configuration Logic

Configuration logic is responsible for determining the role of the node within the system, and for establishing communication routes (internode and intranode). This makes the configuration logic an extremely critical component, and a node cannot function if it is defective. The design uses a hardware test mechanism that exploits transceiver logic to test the configuration block. Since it uses transceiver logic, the test occurs after the transceiver logic test. The test logic first configures the depth-first traversal order of the transceivers within the node, skipping any transceivers that do not generate a "TEST_OK" signal. Next, the test logic uses a 2-bit state machine to circulate a pair of bits (0 and 1) on all virtual channels. If the bits are routed correctly, they arrive back at the insertion point due to the loopback path at the transceivers. If the bits are received correctly, the node generates a CONFIGURATION_OK signal. To avoid masking defects due to defective route setup, each transceiver must ensure that each bit passes through it only once per VC. The configuration test fails if there is a routing error, the bits never return, or the node receives the wrong bit values. A failed configuration test results in a node marked defective.

9.2.2.6 Fail-Stop Compute Logic

Verifying the compute logic in a node is not as critical as verifying the communication and configuration logic. This is because compute logic does not affect system configuration and a node with defective compute logic can be used to improve network connectivity. However, to ensure that the system generates correct results, the compute logic of each node must be tested. This test can be performed at any point before nodes are organized into larger computational entities. This allows the designer the flexibility to implement hardware or software test strategies. In either case, the principle is similar to the previous test strategies—a successful test connects the logic block with the rest of the node; if the test fails, or does not complete, the block remains disconnected from other parts of the node.

Hardware Test

The designer can exploit existing logic to allow repeated execution of test instructions to verify the compute logic. However, this test is unlikely to cover all the logic in the compute block without significant extra hardware. Node size constraints and limited test coverage make this test strategy impractical.

Software Test

Software testing can be performed with minimal additions to the existing node logic. Testing of the compute logic must happen before nodes are organized into larger computational entities. The designer can combine software testing of the compute block with defect isolation by including the test vectors along with the configuration packet. Another advantage of software testing of the compute logic is the possibility of exhaustive testing to ensure correct operation.

The choice of hardware testing for communication and configuration logic, and software testing for compute logic is driven by an analysis of the critical components of a node and technological constraints. As self-assembly technology matures, other test strategies could become more feasible. The next section examines exploiting node modularity to improve system connectivity and tolerate higher transistor defect rates.

9.2.2.7 Using Partially Functional Nodes

The work on NANA, presented in Chapter 7, assumed that a node could either be defective or working correctly. However, if the probability of failure on an individual transistor is high, a larger number of nodes are rendered unusable. The test logic described earlier in this section opens up the possibility of using nodes with some defective components (if they do not affect system operation). For example, a node with a single defective transceiver can still communicate with up to three neighbors and perform computation. The designer can include four modes of failure that allow a node to operate with some defective components, defining each mode based on the number of defects it can tolerate in the compute logic and transceivers. The failure modes are denoted C_xT_y, where x is the maximum number of defects that can be tolerated in compute logic (0 or 1) and y is the maximum number of defective transceivers that can be tolerated (0, 1, 2, or 3). The technique used in the previous chapters cannot tolerate any defects and is denoted C_0T_0. The four additional modes (listed in Table 9.2) are: C_0T_2 (a node cannot tolerate defective compute logic, but can tolerate up to two defective transceivers), C_0T_3, C_1T_2, and a hybrid of C_0T_3 or C_1T_2.

Each failure mode tries to include nodes that could contribute to system operation. The difference is in the minimum operating components each node must have to be used by the system (see Figure 9.2). Nodes are considered useful under C_0T_3 (communication centric) as long as they have one functional transceiver and can be used to compute. Under C_1T_2 (compute centric) a node is useful as long as it has the potential to improve system connectivity by providing an extra path between two parts of the system (i.e., two active transceivers). The hybrid technique includes nodes that can either perform computation, or provide an extra path between two parts of the system. As transistor failure probability increases, the number of nodes marked defective by each technique also increases. Simulations reveal that this increase is fastest for C_0T_0 and slowest for the hybrid failure mode.

Each node requires extra logic to operate with some defective components. This logic keeps track of defective components in the node and disables the node if the defects cross the failure threshold. For example, the C_1T_2 scheme requires 6 bits to keep track of the six primary node components (four transceivers, configuration logic, compute logic). In addition, it requires logic

Table 9.2
Node Failure Modes

Name	Description
C_0T_0	Node can tolerate no failures.
C_0T_2	A node can tolerate up to two defective transceivers (compute logic must work).
C_0T_3	A node can tolerate up to three defective transceivers (compute logic must work).
C_1T_2	A node can tolerate defective compute logic as well as two defective transceivers.
Hybrid	A node can tolerate C_0T_3 or C_1T_2.

C_xT_y defines the number of compute logic (x) and transceiver (y) failures that can be tolerated.

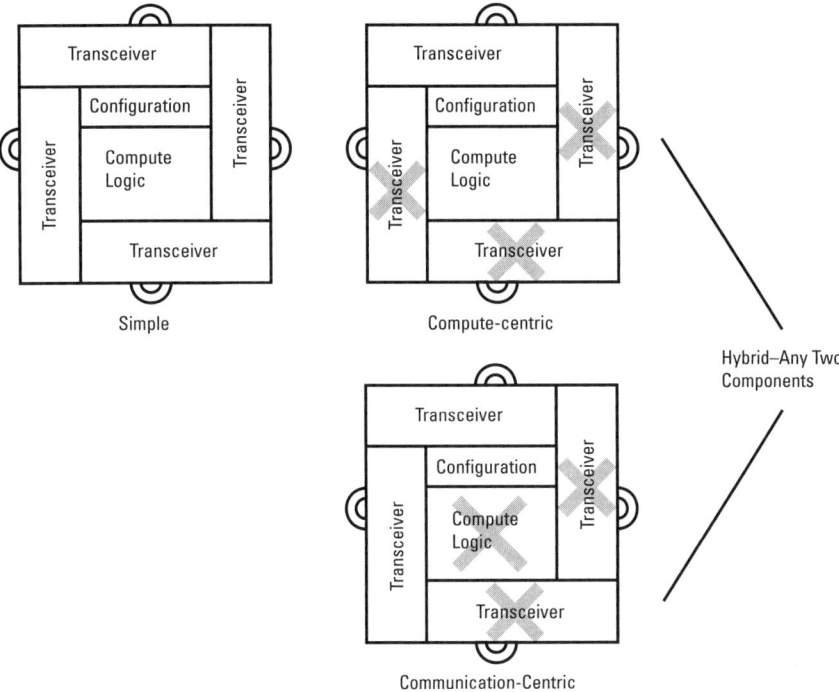

Figure 9.2 Node failure modes.

that determines if more than two transceivers have failed. While this adds to the size of the node, it allows the designer to better utilize each node.

9.2.2.8 Summary

A combination of hardware and software test methodologies is used to verify the operation of a node. The design uses hardware test logic for critical components and relies on software testing for other components. A component can be used only if it undergoes a successful test. This results in fail-stop nodes as defective components are isolated from correctly functioning components. Since the design uses separate tests for node components, with a little extra logic, nodes are allowed to operate even if some (noncritical) components are defective. If transistor reliability is low, allowing these nodes to participate in the system should improve node connectivity. The next section evaluates the effect of using partially defective nodes on the defect isolation mechanism.

9.2.3 Evaluation

This section evaluates three aspects of the proposed fail-stop design. First, it is verified that the test logic for communication and configuration detects defects and measures the overhead of adding the test logic in terms of extra transistors required (Section 9.2.3.1). Next, this section explores the relationship between device failure probability and the expected number of defective nodes in the system, in the context of different node failure modes (Section 9.2.3.2). Finally, this work evaluates the benefit of the testing mechanisms by comparing how well the defect isolation mechanisms perform for different node failure modes (Section 9.2.3.3).

9.2.3.1 Test Logic

The test logic described above is implemented in VHDL and simulated using the Synopsys VHDL debugger. It is first verified that the test circuit generates the TEST_OK signal in the absence of defects in the circuit within a deterministic delay. Next, the response of the test circuit when each signal within the circuit under test is forced to exhibit stuck-at behavior (i.e., forced to 0 or 1) is checked. In each case, it is verified that in the presence of a stuck-at fault, the test logic does not return a TEST_OK signal. Since the test logic circulates a 0 and 1, stuck-at faults can be detected on data lines. Since most data exchanges use handshake signaling, stuck-at faults prevent the circuit from making forward progress (the handshakes require changes in the logic level). The test circuits increase the size of the communication and configuration logic by 18% (736 transistors) and 35% (248 transistors), respectively. The overhead for the configuration logic is higher since the original circuit is not very large.

9.2.3.2 Node Failure Modes

This subsection explores the relationship between the transistor failure probability and defective nodes for different node failure modes. Previous work [2] showed that the reverse path forwarding defect isolation mechanism could tolerate up to 30% defective nodes. That analysis assumed the C_0T_0 failure mode for a node, where 30% defective nodes correspond to a transistor failure probability of less than 4×10^{-5}. It is unclear if self-assembly can guarantee such low transistor failure probabilities. The system can tolerate a higher transistor failure probability by allowing nodes to operate with some defective components. The expected number of defective nodes is computed over a range of transistor failure probabilities, for different failure modes.

A system with 10^6 nodes is used to study the relationship between per-transistor reliability and the fraction of defective nodes. Each node is assumed to have 10,000 transistors, with a uniform device failure probability (P_f). The analysis uses a random number generator (RND) to create values in the interval [0,1]. Each random number corresponds to one transistor in a node. If

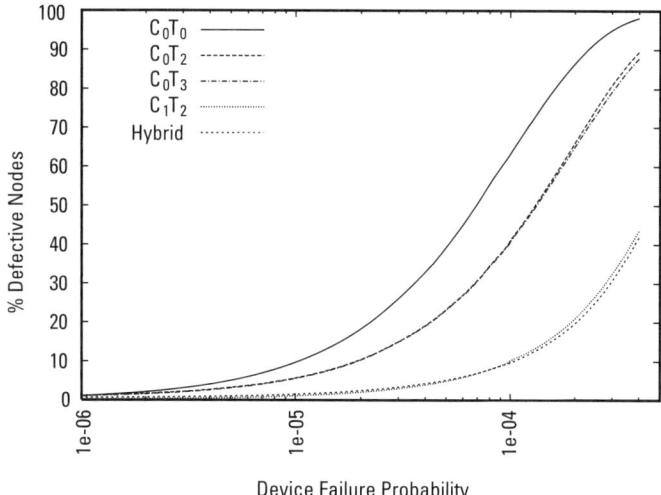

Figure 9.3 Percent defective nodes versus device failure probability for different failure modes.

RND < P_f, the transistor is defective. Each node is then determined to be defective or defect-free for each failure mode. This analysis ignores defective interconnect (within and between nodes). For each value of P_f, 500 experiments are performed with different random seeds.

Figure 9.3 plots the percentage of defective nodes in a system with one million nodes, as a function of the transistor failure probability. Each curve corresponds to one failure mode. As expected, the number of defective nodes in the system decreases as device reliability increases. However, the ability to test components within a node and allow graceful degradation allows the designer to reduce the number of defective nodes without increasing device reliability. It is important to note that for the hybrid failure mode, while a smaller number of nodes are designated defective compared to other failure modes, a large number of nodes have some defective components. While nodes with defective compute logic cannot be used to perform computation, they are useful in improving the connectivity of the network. Next, two baseline network topologies are used to evaluate the benefit of using partially defective nodes.

9.2.3.3 Defect Isolation with Partially Defective Nodes

The previous subsection examined the effect of different node failure modes on the relationship between transistor failure probability and node defect rate. This analysis did not examine the effect of the location of defective nodes on the system. This subsection explores two different node topologies (random and grid) to determine the effectiveness of the defect isolation mechanism with partially defective nodes.

First, the number of nondefective nodes reachable by the broadcast as a function of device failure probability is computed for three node failure modes (C_0T_0, C_0T_3, and hybrid). C_0T_0 is expected to have the lowest number of reachable nodes, followed by C_0T_3, with the hybrid mode having the highest number of reachable nodes. However, a large number of nodes that are reachable with the hybrid mode have defective compute logic and only act towards improving system connectivity. To account for this difference, the number of reachable nodes with operational compute logic is determined.

Figure 9.4 plots the average number of nodes (as a percentage of total nodes) that can be reached for the three failure modes as a function of device failure rate, when nodes are connected in a 100 × 100 grid. For each device failure rate, 100 seed values are used for the random number generator to generate different defect distributions, and compute the average of these 100 runs. Figure 9.4 shows that the hybrid failure mode delivers a significant advantage over C_0T_0 and C_0T_3 (even if only nodes with functioning compute logic are considered). While there is a sharp decrease in the number of reachable nodes beyond a certain device defect probability, this threshold is higher with hybrid failure than with C_0T_0 failure.

The number of nondefective nodes that are reachable in a random network with 10,000 nodes is also computed. This random network is meant

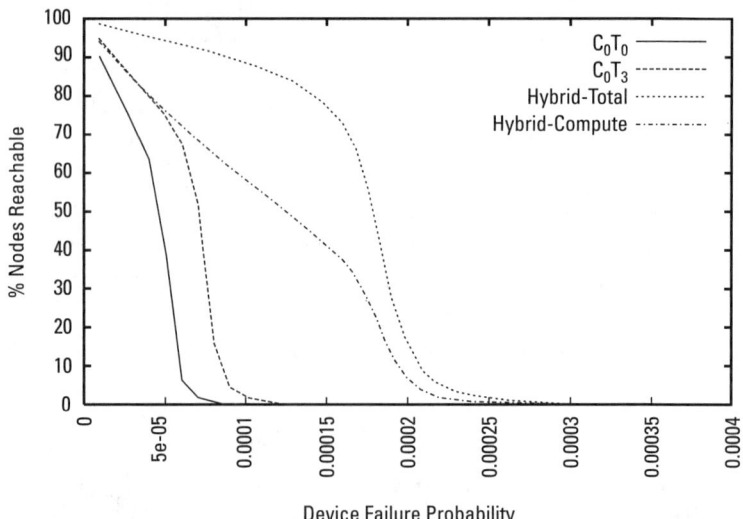

Figure 9.4 Percent nodes reachable versus device failure probability for a grid with different node failure modes.

Overcoming Randomness with Increased Complexity

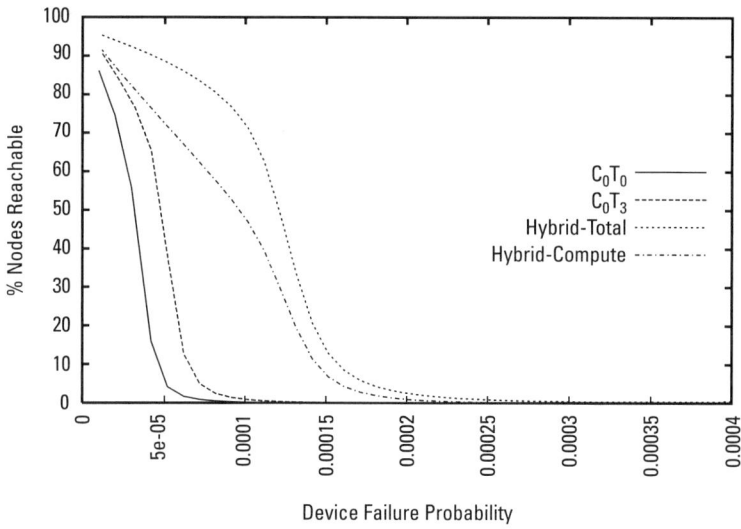

Figure 9.5 Percent reachable nodes versus device failure probability for a random network with different node failure modes.

to be representative of self-assembled networks of nodes. The random network has inherently lower connectivity than a regular grid and some nodes might be disconnected from the rest of the network. This implies that there should be a reduction in the failure probability threshold for all schemes. For the random networks, 100 random topologies are generated, and then 100 seed values are used per topology to get statistically accurate results. Figure 9.5 plots the average number of nodes (as a percentage of total nodes) that can be reached for the three failure modes. As expected, the knees in the curves have shifted left, but the general shapes are similar to those seen for a regular grid.

Finally, this work evaluates the benefit of using the hybrid failure mode over C_0T_0. Figure 9.6 plots the average fraction of all nodes that are reachable as a function of the percentage of defective nodes as defined by C_0T_0 (i.e., single defect renders node unusable). There are two curves each for two types of network topologies (grid and random). The two curves correspond to the number of nodes reachable using C_0T_0 and the number of nodes with functioning compute logic reachable when using the hybrid failure mode. Note that the total number of nodes reachable by the hybrid mode is greater than those reachable with functioning compute logic, since the hybrid mode uses nodes with defective compute logic and two (or more) functioning transceivers. The hybrid failure mode allows the system to include nodes that would be unusable with C_0T_0.

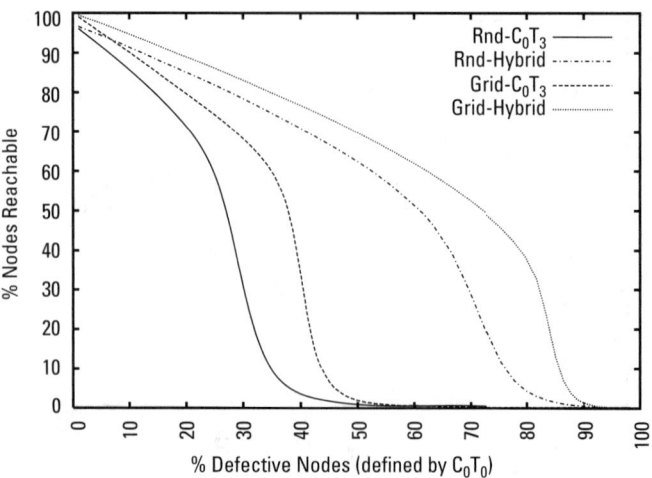

Figure 9.6 Effect of partial node failure.

9.2.3.4 Result Summary

The above results show that allowing partially defective nodes to participate in system operation increases the transistor failure probability that can be tolerated by the system. Allowing nodes with defective compute logic but functional communication logic to remain in the system improves network connectivity.

9.2.4 Related Work

There has been extensive work on testing circuits. The key difference in the work described above is the scale of the nodes, and the technological constraints that limit circuit size. Built-in self-test (BIST) is a common technique used to test circuits. BIST strategies typically use linear feedback shift registers (LFSRs) to generate pseudo-random test patterns for circuits. While most BIST strategies are used with synchronous circuits, there are asynchronous BIST techniques as well [4, 5]. Both the hardware test mechanisms presented in this chapter are a form of BIST. The design does not require (and cannot fit) large LFSRs since it targets simple circuits and single-bit wide data paths.

The Teramac custom computer [6] from HP Labs is built out of a large set of partially defective FPGAs. It uses externally initiated testing to obtain an external defect map of the FPGAs. It then configures the system to isolate defective regions. The proposed system is significantly larger than the Teramac, making defect map extraction infeasible.

9.2.5 Summary

This section presents a scheme for achieving fail-stop behavior in limited size nodes by dividing them into modular components. It analyzes the trade-offs in implementing hardware/software test schemes for the components, and the use hardware testing for critical node logic and software testing for other logic. The use of partially functional nodes improves network connectivity, and helps the system tolerate devices with higher failure probabilities (increased from 4×10^{-5} to 1.5×10^{-4}). As self-assembly matures as a technology, node size restrictions could decrease, allowing the use of faster and more comprehensive hardware test schemes.

9.3 Self-Assembled Networks: Control Versus Complexity

The fail-stop designs presented in the previous section can help overcome defects introduced by randomness during the fabrication of a single node (i.e., defects in the circuit of a node). Furthermore, the built-in self-test circuits for the transceivers can detect when a physical link is not connected to any other node—a fabrication defect introduced during the second phase of self-assembly. In addition to internode connections, randomness during the second phase of assembly influences node placement and node orientation.

The previous discussions of NANA (Chapter 7) and SOSA (Chapter 8) assume either a mesh or a random network of nodes. However, these two topologies lie at opposite ends of a spectrum defined by the amount of control exercised over the self-assembly process. This section examines network properties for varying degrees of control over how nodes are placed and oriented, and how internode links are created during self-assembly. This enables examining a range of networks, from a mesh (full control) to a random network of nodes (no control). For each network type, network connectivity is determined along with the need for any additional hardware in each node's communication logic to maximize the number of connected nodes. Simulation of the SOSA data parallel architecture built on top of these networks running a simple benchmark is used to evaluate the system performance, which is shown to be independent of the underlying network, as long as sufficient nodes are available for computation. Finally, it is observed that the introduction of defects in nodes and links can exacerbate the poor connectivity found in networks with low control during self-assembly.

The rest of this subsection is organized as follows. First, it provides a brief overview of the node and system architecture. Next, it describes the different networks that are created versus various levels of control over self-assembly. Finally, it presents an evaluation of network connectivity with and without defects, and the impact on system performance.

9.3.1 Node and System Architecture

The SOSA system (Chapter 8) forms the basis for this work and thus only the relevant details are summarized here.

9.3.1.1 Node Architecture

Each node has three primary components: (1) communication logic, (2) configuration logic, and (3) compute logic. The communication logic has four transceivers that allow the node to communicate with other nodes over single wire links. In case more than two transceivers share a single wire, the baseline transceiver model implements an infinite backoff mechanism that permits only two active transceivers on that link. When the node is powered up, each component undergoes a simple test to ensure correct operation as described in the previous section. If the transceivers pass this test, they attempt to signal neighboring nodes over their wire link. If a transceiver detects more than one other transceiver signal over the link, it shuts down. This can potentially affect network connectivity in cases where the transceiver that shuts down provides access to a region of the network.

9.3.1.2 System Architecture

To maximize node utilization, SOSA is a data parallel architecture implemented on the network of nodes. Since a single node does not have the necessary compute power, sets of nodes are logically grouped together to form SIMD style processing elements (PEs) connected in a logical ring. This system architecture supports a data parallel programming model. The topology of the underlying network affects the logical organization of PEs in the logical ring, which in turn can have an effect on system performance. This potential performance impact is explored by analyzing the performance of one data parallel program on various networks.

9.3.2 Networks of Self-Assembled Processing Elements

The node network topology depends on the level of control exercised during node self-assembly, and during the creation of internode links. As self-assembly technology matures, it might be possible to create three-dimensional topologies as well, but this study is limited to the analysis of two-dimensional topologies. The explored topologies are created by varying control over three aspects of the manufacturing process:

- Placement of nodes (P);
- Orientation of nodes (O);
- Creation of internode links (I).

In each case, to limit the parameter space to be explored, two alternatives are considered: (1) full control, and (2) no control. This results in eight network types, ranging from a random planar network to a mesh. Table 9.3 lists the networks by the type of control necessary to create them and Figure 9.7 shows examples of these networks. The goal is to identify the level of control necessary to maximize the number of connected nodes. To this end, the following subsections discuss techniques for potentially controlling placement, orientation or internode link creation, and the implications of that control on the number of connected nodes in the network.

9.3.2.1 Placement (P)

Control over node placement enables uniformly spaced nodes. It is expected that uniform spacing improves network connectivity by reducing the number of nodes that require long links to connect to the rest of the network. Control over node placement can be achieved in two ways: (1) pick and place techniques, and (2) placing DNA tags on the underlying substrate to control the locations where nodes self-assemble. While each node is large enough to enable the use of pick and place strategies, they are not practical for systems

Table 9.3
Classification of Network Topologies Based on Control over Placement, Orientation, and Interconnect

Name	Control			Example
	Placement(P)	Orientation(O)	Link(I)	
N0	No	No	No	Figure 9.7(a)
N1	No	No	Yes	Figure 9.7(b)
N2	No	Yes	No	Figure 9.7(c)
N3	No	Yes	Yes	Figure 9.7(d)
N4	Yes	No	No	Figure 9.7(e)
N5	Yes	No	Yes	Figure 9.7(f)
N6	Yes	Yes	No	Figure 9.7(g)
N7	Yes	Yes	Yes	Figure 9.7(h)

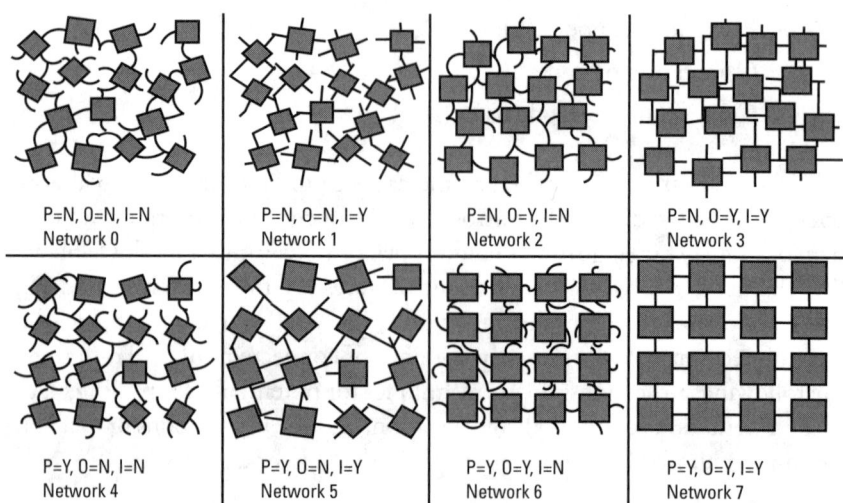

Figure 9.7 Examples of eight networks with varying control over placement (P), orientation (O), and internode link creation (I). (a–d) No control over P; nodes can get isolated due to large distances, control over O and I improve connectivity. (e–h) Control over P improves connectivity, but can still result in isolated nodes without control over O and I.

with a large number of nodes. System designers can minimize external intervention by placing DNA tags on the substrate to initiate node growth at tag locations. The designer can increase the chance that a node forms at the correct location by using more tags per node. However, increasing the number of tags increases the effort required in preparing the substrate for self-assembly. Examples of the network types created with node placement can be found in Figure 9.7(e–h).

9.3.2.2 Orientation (O)

Control over orientation aligns node faces, which can increase the chances of links intersecting—as depicted in Figure 9.7(c, d, g, h)—potentially improving network connectivity. The techniques to control node placement could also be extended to control node orientation by increasing the number of tags per node. In addition to using multiple tags on the substrate, nodes could be aligned using an external electric field, or by using fluid flow [7].

9.3.2.3 Internode Link Creation (I)

Control over internode link creation implies control over the shape of links. Without creating a mesh network (and linear links), there is still a chance that more than two transceivers are connected by a link. Linear links cannot loop back on themselves and are useful in improving network connectivity. Researchers have demonstrated the creation of mostly linear wire structures [8, 9]. Networks with linear links are shown in Figure 9.7(b, d, f, h).

9.3.2.4 Maximizing Reachable Nodes

In any nonmesh topology, more than two links (and transceivers) can potentially be connected. The baseline node design deals with such links by implementing an infinite backoff mechanism, which essentially disables the transceiver. However, disabling a transceiver could potentially partition the network. This problem can be mitigated by treating each link like a shared medium (bus). However, this requires extending transceiver functionality to enable arbitration for link access, as well as the use of source/destination identifiers for each transfer on the link. This study evaluates the potential benefit of one method for link sharing. As self-assembly matures, it might be possible to create larger nodes that incorporate this extra functionality.

9.3.3 Experimental Setup and Evaluation

The evaluation begins with a description of the network topology generator and the methodology used to model infinite backoff and link sharing between transceivers (Section 9.3.3.1). The goal of this evaluation is to study the connectivity characteristics of each network type and determine if the baseline node design needs to be augmented to maintain good system connectivity (Section 9.3.3.2). It is also desirable to determine the impact of the different networks on system performance (Section 9.3.3.3). Finally, this work studies the effect of node and link defects on connectivity (Section 9.3.3.4).

9.3.3.1 Topology Generator

The topology generator's inputs include the number of nodes, total area, type of control over placement (P), orientation (O), interconnect (I), minimum distance between nodes, and an optional parameter that decelerates interconnect growth with time. It also accepts a random seed which allows the creation of distinct topologies. For networks with no control over node placement (P = N in Figure 9.7), it generates a random location for the node and places it there if all constraints are met (no overlap, minimum distance, within area). The program attempts to place each node a maximum of 10^6 times. For networks with control over node placement (P = Y in Figure 9.7), a simple check of the area and number of nodes allows the program to determine if the nodes fit. If O = N, each node is rotated (about its center) through a random angle before being placed.

After placing all nodes, the simulator models link growth. For random growth (I = N in Figure 9.7) the simulator uses a random number generator and a probability distribution function (PDF) for the angle and distance by which the link grows to perform a directed random walk. If linear growth (I = Y in Figure 9.7) is modeled, the link is extended by a random length (< = 50 nm). Each link is iteratively grown by this random length until one of two conditions is satisfied: (1) it collides with another node or link, or (2) the simulation terminates as a user-defined condition is satisfied. Once growth of all links terminates, the simulator generates a graph corresponding to the node network created by the links and generates connectivity statistics for the graph.

Modeling Infinite Backoff

To model infinite backoff, the simulator identifies links with more than two transceivers, randomly picks two transceivers to be active, and disconnects the rest. There are multiple ways of picking a pair of transceivers and the simulator can generate multiple networks by randomly picking different pairs of transceivers.

Modeling Links as Buses.

The simulator models one possible implementation of shared links, where the N transceivers connected by a single link are divided into pairs that communicate with each other. If N is odd, one transceiver is not used. There are multiple ways in which the transceivers can be paired and this effect is captured by creating multiple networks with different transceiver pairs.

9.3.3.2 Reachable Nodes

For each of eight network types, 100 topologies for different network sizes are generated. The network sizes for this study are 10,000 nodes, and the number of nodes required for the smallest square mesh required to run 8 × 8 (1,296 nodes), 16 × 16 (4,900 nodes), and 32 × 32 (21,025 nodes) matrix multiplication.

Figure 9.8 plots the size of the largest connected group of nodes as a fraction of total nodes for networks with 21,025 nodes (other network sizes show similar results). For each network type there are three bars, the first representing the unconstrained network, the second corresponding to links modeled as buses, and the third with transceivers implementing infinite backoff. Figure 9.8 shows that if connectivity is unconstrained, all networks are able to connect in excess of 95% of the nodes. However, when modeling realistic hardware the fraction of reachable nodes decreases. The decrease is small when links are treated as shared media. However, with infinite backoff on links, for networks without control over placement and orientation, less than 50% nodes are reachable.

The average number of transceivers connected per link is shown in Figure 9.9. For a fully connected network of nodes there would be two transceivers per link and 1.97 for a mesh since the boundary transceivers are disconnected. For the unconstrained networks the value is more than 2, indicating that multiple transceivers share links. The value drops to about 1.7 for shared links and about 1.55 for links with infinite backoff. This implies that

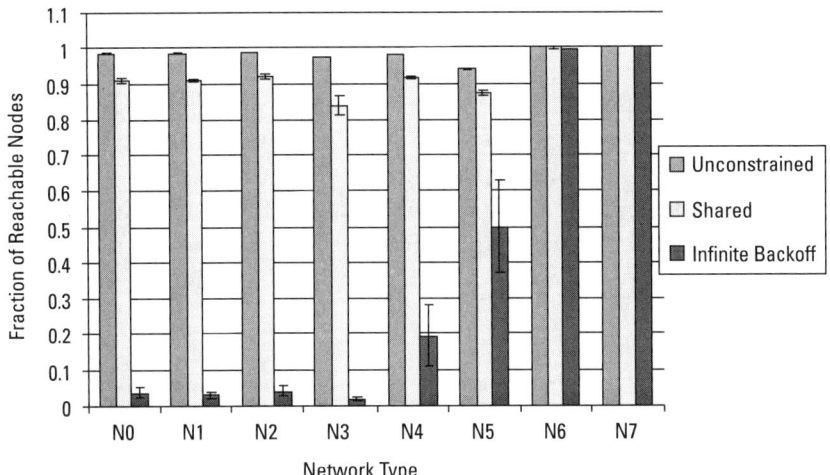

Figure 9.8 Fraction of reachable nodes.

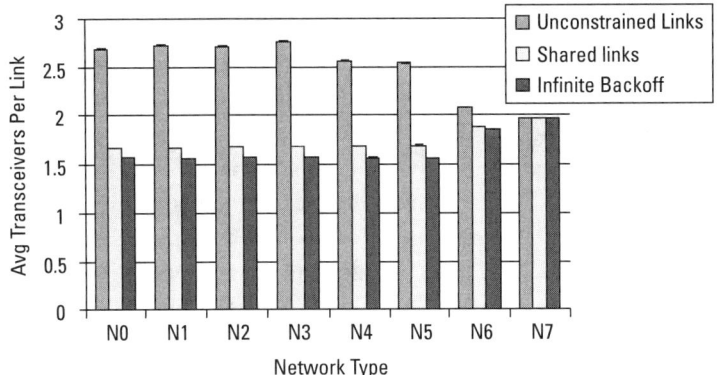

Figure 9.9 Transceivers per link.

only 55% of all links are connected to a second transceiver if it implements infinite backoff, which explains the poor network connectivity.

This highlights the trade-off between simple nodes and the degree of control required during self-assembly to achieve good network connectivity. Simpler nodes require regular topologies to achieve good connectivity. If nodes can implement mechanisms to allow more than two transceivers to share a single link, the system is well connected even if there is no control over the manufacturing process.

9.3.3.3 System Performance

In an ideal system, performance would be independent of network topology. The effect of topology on system performance is quantified by measuring the running time of an application (matrix multiplication) on different networks using a simualtor for the SOSA data parallel architecture. Program run-time is measured for networks with at most two transceivers sharing links (infinite backoff), or pairs of transceivers sharing links (links as buses). This study simulates matrix multiplication for three matrix sizes: 8×8, 16×16, and 32×32. The results reveal that as long as enough PEs can be configured in the network, there is very little variation in program running time.

9.3.3.4 Effect of Defects

To study the effect of defects on network connectivity, this work uses the hybrid node failure model from the previous section with a range of device reliabilities. Table 9.4 lists the percentage of nodes that are reachable in a network of 21,025 nodes for shared links. The numbers in parentheses are the percentage of reachable nodes when modeling infinite backoff. This number is omitted if it is less than 10%. These results show that the percentage of reachable nodes drops rapidly as device reliability or control over self-assembly decreases. System connectivity decreases since some regions get disconnected due to the loss of critical nodes/links to defects. This is reflected by a drop in the number of transceivers per link (between 22% to 50% drop) as the device reliability decreases from 100% to 99.99%.

The results highlight the benefit of link sharing over infinite backoff. Link sharing allows a larger number of nodes to remain connected as device reliability decreases. This is true even for configurations with low control during self-assembly (N0-N3). These results lead to two conclusions: (1) if device reliability is lower than 99.999%, the designer either needs to control placement and orientation during self-assembly, or the designer needs to implement link sharing to maintain network connectivity; and (2) controlling placement and orientation has a greater effect on network connectivity than link sharing. The two techniques can be combined to achieve greater effect.

Table 9.4
Percentage of Nodes Reachable with Varying Device Reliabilities when Links Are Shared Between Multiple Trasceivers

Device Reliability (%)	Configuration							
	N0	N1	N2	N3	N4	N5	N6	N7
99.990	3.8	4.9	3.9	2.5	3.4	6.6	86 (84)	90 (90)
99.993	18	26	19	7.2	18	38	91 (91)	93 (93)
99.996	73	73	73	31	72	72	95 (95)	96 (96)
99.999	88	88	89	78	88.6 (11.7)	84.7 (35.5)	99 (99)	99 (99)
100.00	91	91	92	84	92 (19.5)	87 (50)	100 (100)	100 (100)

The value in parantheses is for nodes implementing infinite backoff

9.3.4 Summary

This section evaluates the characteristics of a class of network topologies that could be created by exercising varying degrees of control during the self-assembly of simple nodes. The evaluation highlights the trade-off between node complexity and the amount of control required during self-assembly to maximize the number of connected nodes in the network. This study also reveals that if the network has enough nodes, system performance is not affected by the type of configuration created by self-assembly. Finally, it is observed that introducing defects has a greater effect on networks with a lower degree of control during self-assembly. However, this can be mitigated to some extent by allowing more than two transceivers to share a link.

References

[1] Dalal, Y. K., and R. M. Metcalfe, "Reverse Path Forwarding of Broadcast Packets," *Communications of the ACM*, Vol. 21, No. 12, 1978, pp. 1040–1048.

[2] Patwardhan, J. P., et al., "Evaluating the Connectivity of Self-Assembled Networks of Nanoscale Processing Elements," *Proc. IEEE International Workshop on Design and Test of Defect-Tolerant Nanoscale Architectures (NANOARCH '05)*, 2005, pp. 2.1–2.8.

[3]   Dally, W. J., "Virtual-Channel Flow Control," *Proc. 17th Annual International Symposium on Computer Architecture*, 1990, pp. 60–68.

[4]   Petlin, O. A., and S. B. Furber, "Built-In Self-Testing of Micropipelines," *Proc. 3rd International Symposium on Advanced Research in Asynchronous Circuits and Systems*, 1997, p. 22.

[5]   Alves, V. C., F. M. G. Franca, and E. P. Granja, "A BIST Scheme for Asynchronous Logic," *Proc. 7th Asian Test Symposium*, 1998, p. 27.

[6]   Culbertson, W. B., et al., "The Teramac Custom Computer: Extending the Limits with Defect Tolerance," *Proc. IEEE International Symposium on Defect and Fault Tolerance in VLSI Systems*, 1996, pp. 2–10.

[7]   Huang, Y., et al., "Directed Assembly of One-Dimensional Nanostructures into Functional Networks," *Science*, Vol. 291, 2001, pp. 630–633.

[8]   Kudo, H., and M. Fujihira, "DNA-Templated Copper Nanowire Fabrication by a Two-Step Process Involving Electroless Metallization," *IEEE Transactions on Nanotechnology*, Vol. 5, No. 2, 2006, pp. 90–92.

[9]   Fukunaka, Y., et al., "Producing Shape-Controlled Metal Nanowires and Nanotubes by an Electrochemical Method," *Electrochemical and Solid-State Letters*, Vol. 9, No. 3, 2006, pp. C62–C64.

Appendix: Laboratory Methods in DNA Self-Assembly

This appendix describes how to create DNA self-assembled nanostructures starting from the commercially available raw materials. Included in this appendix are some general guidelines common in the field which have been collected over time from experience and from discussions with other groups. This appendix is intended to serve as an introduction to those who are interested in DNA self-assembly laboratory experiments but who have little experience in a wet (biochemistry) lab.

A.1 General Lab Practices

There are many factors that can perturb laboratory work and negatively influence the quality of experimental data. Sample contamination, pipetting errors, and improper preparation of ancillary materials like AFM tips and mica are just a few. Such problems can be avoided by following a simple set of general lab practices.

Clean Working Environment. DNA structures are sensitive to environmental contamination and the impact of contaminants range from subtle conformational changes to full denaturation of secondary structures and even breakdown of the DNA backbone (e.g., when DNA nucleases are present). DNA nucleases are very common (e.g., in the oils on our skin for instance), but they can be kept separate from experimental samples by maintaining a clean working environment. Some guidelines are offered here:

1. Always wear protective equipment while in a lab. At a minimum this includes gloves and a lab coat. It is not only good practice but also a safety requirement: it protects both you and your experimental samples.
2. Use a fume hood or a sealed chamber when manipulating potential contaminants.

3. Keep your workbench clean so that any accidental spills can be easily noticed and dealt with properly.
4. Do not bring any food or drink into the lab since these can carry many varieties of contaminants and can potentially be contaminated.

Particulate Contamination and Filtering. Chemical reagents often include residual particles or fibers from their synthesis that can be contaminants. Water and buffer materials are particularly susceptible to this effect. Use a vacuum filter with a small-pore membrane to filter such chemicals before use. Airborne dust can also become a contaminant if your lab is not located inside a cleanroom.

Pipetting. Pipetting-induced contamination is usually due to inadvertent tip reuse across samples. This can be avoided by always using a new tip for each pipetting action. While there are cases where tip reuse does not have a negative impact and one could save a few tips, always using a new tip creates a good habit which works to prevent this type of contamination. This can be very useful when long pipetting sequences are required.

DNA and Surfaces. When using AFM to image DNA structures samples come into contact with mica and tip surfaces. Proper preparation of these surfaces avoids contamination and helps imaging. Since AFM relies on the sample-surface contact adhesion it is important to make reliable and repeatable surface conditions.

- *Mica.* Use UV-curing glue to bind mica to metallic AFM pucks. Add just enough glue so that the entire mica-metal interface is filled but that none overflows the edges. Do not press on the mica to force this process since this will only damage the mica surface and possibly layers beneath. Create a fresh cleave (using tape or tweezers) each time you need to deposit a sample for imaging. Visually inspect it to ensure that the cleaving action resulted in a single, clean plane of mica (i.e., there should be no color variation or interference patterns on the surface). During deposition ensure that enough sample (i.e., liquid) is present so that evaporative loss during imaging does not dry out the mica surface.
- *AFM tips.* Use O_2 plasma to clean tips for 5 minutes and rinse in ddH_2O (double distilled, de-ionized water). To prevent DNA from sticking to the tip surface, use a 1 mg/mL solution of polyethylene glycol (PEG, ~10 kD) in ddH_2O when rinsing. This creates a protective PEG layer which is particularly useful for imaging DNA/protein

samples. The converse is true if using PEG on a flat surface (e.g., mica or silicon); a surface layer of small chain length PEG will *promote* DNA adhesion to the surface.

A.2 Creation and Storage of Buffers

The commonly used buffer for DNA self-assembly is a tris-acetate-EDTA. To generate a working buffer from stock materials a higher concentration of buffer must be used. The role of magnesium here is two-fold: (1) to serve as the counter ion for DNA, and (2) to aid the deposition process of DNA onto mica for AFM imaging. In all cases, a counter ion is required for DNA to be highly soluble in water. For more general preparations, the buffer compound (e.g., TAE) is sufficient to solubilize the DNA. The following procedure describes the process for making this buffer.

Procedure 1
Makes 1 Liter of 10X TAE Mg^{2+} stock (125 mM Mg Acetate)

Materials:	*Mass or Volume*
Hydrous Magnesium Acetate	26.8g
50X TAE stock	200 mL
ddH_2O	to 1,000 mL

Procedures:
1. Add solid Mg Acetate 4H_2O.
2. Add 50X TAE stock.
3. Fill to ~600 ml mark with Milli-Q H_2O.
4. Nuke for 1–2 minutes (loosely stoppered) in a warm water bath of 40–50°C.
5. Mix until the Mg is dissolved.
6. Add milli-Q to 1 liter mark.
7. Filter the buffer with a 0.2-micron filter.*
8. Store the buffer at 4°C in the refrigerator.

To modify the pH, add acetic acid to the buffer while monitoring the result with a pH meter. The recommended pH range for DNA self-assembly is 7.5.

NOTE: Filtration of the buffer is done using a vacuum filter apparatus. The setup is composed of several components: the container, the filter mount, the vacuum chamber, and the flask for sample collection. The filter is a polycarbonate disc and should be placed over the filter mount which fits into the ring on the vacuum chamber. A clamp is used to hold the vacuum chamber and the container for the buffer (a typical volume for a small setup is 250 mL). Filtration usually takes no more than 5 minutes.

A.3 Aliquotting DNA Strands

Uniformity in concentrations between DNA samples will help simplify procedures. Thus, it is important to aliquot carefully since it is a time-consuming and tedious process (to repeat) and an important first step. The following procedures describe how to aliquot, store, and verify the presence of DNA in the aliquot starting from raw single-stranded DNA.

The standard procedure, provided here as a guideline *only*, is derived for the final concentration of tiles to be approximately 1 (M. The arms and cores must both be diluted to 8 (M, while the shells must be diluted to 24 (M. These are simply convenient concentrations and can be changed. If you do need to adjust the concentration, remember to keep the final arm:core:shell ratio at 4:1:4.

Strands either arrive from the company as lyophilized (dried) powder or already suspended in standard buffer. In both cases, the volume needed to dilute the strands to their appropriate concentrations must be determined. The following information is needed from the data sheet for these calculations:

- For powder form: (1) number of moles (usually in nmoles), and (2) extinction coefficient at 260nm;
- For strands in buffer: (1) number of moles, (2) total volume present, and (3) extinction coefficient at 260nm.

To determine the volume of 1X TAE buffer needed, simply divide the number of moles by the number of moles for the final concentration. For example, for a lyophilized powder of an arm strand with 9 nmoles:

- 9 nmoles/8 nmoles/mL = 1.125 mL 1X TAE buffer needed.
- Check: 8 nmoles/mL = 8 (moles/L = 8 (M final concentration.

For strands that are suspended in buffer, the procedure is the same except that the volume currently present in the tube must be subtracted from the added volume. For example, the same arm strand in 125 (L of buffer needs:

- 9 nmoles/8 nmoles/mL − 0.125 mL = 1.0 mL 1X TAE buffer to be added

After adding the needed volume, use the following procedure to ensure a well-mixed volume:

1. Mix well for about 1 minute.*
2. Allow the tube to stand for 5 to 10 minutes.
3. Mix again for 1 minute.
4. Vortex and spin (centrifuge to remove any sample from the tube's lid).

*Mixing procedures can vary. The suggested technique is to generally mix a number of times (10–40 times) by pipetting repeatedly, using a volume which is 33% to 50% of the total volume being mixed to ensure homogeneity.

Once the DNA has been sufficiently mixed it can be aliquotted (i.e., divided into equal parts) into a specified volume (e.g., 150 (L).

It is ideal to label each aliquot by the aliquot number and the volume contained and to include information from the data sheet like the date created, the assumed concentration, and the ID number. All aliquots (save the first aliquot) should be stored at −80°C to reduce the impact of hydrolysis on the DNA that occurs at temperatures ≤80°C. The first aliquot should be used to double-check the achieved concentration. Dilute the volume with buffer to a 1:20 ratio for measuring in a UV-Vis spectrometer. The peak should be at approximately 260 nm and the ratio of the A260 vs. A280 readings should be close to 1.8. Using Beer's law ($A = \alpha lc$) calculate the concentration from the peak. If done properly the measured concentration should be within 5% of the expected concentration. Adjust as needed and record this information for subsequent aliquots. The 1:20 dilution (and mixing!) should occur *in the cuvette* after recording a baseline of the diluting buffer. The mechanical replacement error (i.e., taking a cuvette out of the spectrometer and replacing it) can vary by instrument, day, ambient temperature, etc. and induces noise in the measurement that can exceed 5%.

Note that chemical modifications to the DNA may create slight differences in the absorbance spectra. However, as long as the absorbance at 260 nm remains unperturbed this procedure will work.

Using the guidelines above Table A.1 shows the uncorrected volumes, assuming precise concentrations of the strands, components and quantities to anneal tiles.

Table A.1
Volumes of DNA for the 16-tile Grid

		Volume
Cores and arms (1 core and 4 arms)	7.5 µL each	37.5 µL
Shells (4 shells)	2.5 µL each	10 µL
10X TAE Mg2+	6 µL	6 µL
Milli-Q H_2O	26 µL	6.5 µL
Total volume:		60 µL

If strands are not aliquotted to precise concentrations, then corrections factors need to be applied to the procedure to account for the actual concentration of each strand. Once the volumes are corrected, the sum of the volumes for each strand type as well as the buffer can be subtracted from 60 (L (i.e., the working volume). The remaining volume is the volume of ddH_2O that is needed to complete the procedure.

A.4 Tile Annealing

DNA tiles are structural building blocks for more complex structures (e.g., grids). They are formed through the hybridization of their constituent single strands of DNA: one core, four (distinct) shells, and four (distinct) arms. The sequences of these strands are designed so that when they hybridize they form a cruciform tile structure, as shown in Figure A.1.

Add the component strands for each tile into a centrifuge tube and thoroughly mix using a pipette set for half the total sample volume (30 uL for typical 60-uL tile volumes). For 4 × 4 fully addressable DNA grids, each of the 16 tiles has unique arms and reused core/shell strands. Use premade procedure tables that indicate the precise strands and quantities used for each tile.

A temperature-controlled water bath is used for the annealing process. To ensure good thermal mixing of the constituent strands, the annealing starts at a high 95°C. The temperature linearly drops to 4°C over a period of 24 hours and is then kept at a constant 4°C for 4 hours. Vortex and spin the tubes, then store tiles at 4°C for up to 6 weeks.

Note, each centrifuge tube containing a tile should have their tops wrapped in a strip of paraffin to prevent the tube from popping open during the annealing process.

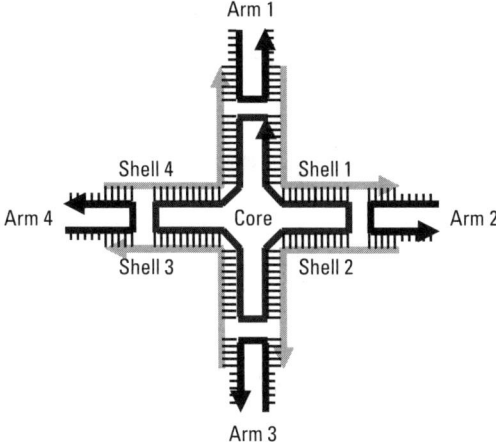

Figure A.1 Cruciform tile: one core, four shells, and four arms.

A.5 Grid Annealing

Although the structural domain of a tile largely consists of hybridized double-stranded DNA, 10 bases at the ends of each arm remain unpaired. These single-stranded sequences are the sticky-ends that allow the formation of grid structures. When sticky-ends from different tiles have complementary sequences, they tend to hybridize and in that process bind tiles together. Through the sequence design of sticky-ends, the specific position of each tile on the grid can be engineered to be thermodynamically favored. The 4 × 4 fully addressable DNA grid has 16 unique tiles, as shown in Figure A.2.

To anneal 4 × 4 grids, mix equal amounts of each tile in a centrifuge tube. A water bath can be used to anneal the sample at a constant 23°C over a 4-hour period. Cool the sample to 4°C and keep at 4°C for 4 hours. Vortex and spin the tubes, and store tiles at 4°C for up to 4 weeks. Grids can survive as long as 12 months at 4°C. However, this relatively long storage time induces noticeable degradation of most grid structures.

Table A.2 lists the DNA sequences and strand names for the 16-tile grid.

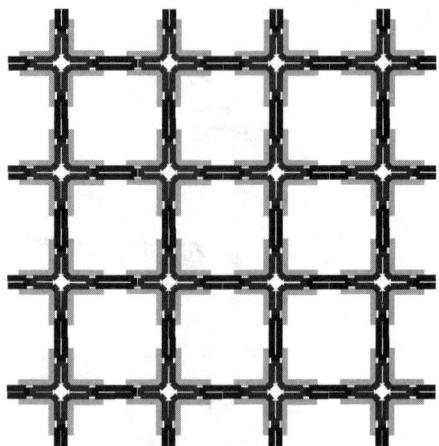

Figure A.2 The 4 × 4 fully addressable grid with 16 unique tiles.

Table A.2
DNA Sequences, Synthesis Scale, and Purification Method for the 16-tile Grid

Strand	Sequence	Scale	Purification
core A	aggcaccatcgtaggttttcgttgcgatca ccaacggagttttttctgccgtacaccagtgaag tttttcgatcctagcacctctggagttttcttgcc	1 umole	PAGE
core B	taacaccttcgctcgttttcgagtgagacaccgccgacc tttttgacgagaccacctatcatctttt tcgatagtgtcaccggccttcttttcgacgg	1 umole	PAGE
shell A-1	atgcaacctgcctggcaagactccagaggac tactcatccgt	250 nmole	HPLC
shell A-2	tccgactgagccctgctaggatcgacttcactg gaccgttctaccga	250 nmole	HPLC
shell A-3	accggaggcttcctgtacggcagaactccgttg gacgaacag	250 nmole	HPLC
shell A-4	atagcgcctgatcggaacgcctacgatggacacgccg	250 nmole	HPLC
shell B-1	gattaccctgttaccgtcgagaaggccggacatca gttcagc	250 nmole	HPLC
shell B-2	gaaccgcgaatcctgacactatcgagatgataggac aacgccatacc	250 nmole	HPLC
shell B-3	atcgacagatccctggtctcgtcaaggtcggcggactctatc	250 nmole	HPLC
shell B-4	gatgtacctgtctcactcgcgagcgaaggacgctacc	250 nmole	HPLC
arm 1.1	gttatcggcgtgtggttgcataatac	100 nmole	HPLC
arm 1.2	caatcacggatgagtagtgggctcagtcggacattc	100 nmole	HPLC
arm 1.3	cctcgtcggtagaacggtggaagcctccggtcgtgc	100 nmole	HPLC
arm 1.4	ttcaactgttcgtggcgctatattgt	100 nmole	HPLC
arm 2.1	caagccggcgtgtggttgcatacgac	100 nmole	HPLC
arm 2.2	aagtgacggatgagtagtgggctcagtcggatactg	100 nmole	HPLC
arm 2.3	ttgattcggtagaacggtggaagcctccggtttaca	100 nmole	HPLC
arm 2.4	gattgctgttcgtggcgctatgaatg	100 nmole	HPLC
arm 3.1	ttaacggtagcgtgggtaatctatga	100 nmole	HPLC
arm 3.2	aatgcgctgaactgatgtggattcgcggttcattgc	100 nmole	HPLC
arm 3.3	cagacggtatggcgttgtgggatctgtcgatacgtc	100 nmole	HPLC
arm 3.4	cacttgatagagtggtacatccagta	100 nmole	HPLC
arm 4.1	acagcggtagcgtgggtaatcactgc	100 nmole	HPLC
arm 4.2	caatagctgaactgatgtggattcgcggttctaatg	100 nmole	HPLC
arm 4.3	tacttggtatggcgttgtgggatctgtcgatttact	100 nmole	HPLC

(*continues*)

Table A.2
Continued

Strand	Sequence	Scale	Purification
arm 4.4	gcattgatagagtggtacatcgcaat	100 nmole	HPLC
arm 5.1	cgaggcggcgtgtggttgcatgcacg	100 nmole	HPLC
arm 5.2	ttaagacggatgagtagtgggctcagtcggattgta	100 nmole	HPLC
arm 5.3	tcatgtcggtagaacggtggaagcctccggttttgct	100 nmole	HPLC
arm 5.4	tgtagctgttcgtggcgctattacgt	100 nmole	HPLC
arm 6.1	tctgacggcgtgtggttgcattcaac	100 nmole	HPLC
arm 6.2	ctacaacggatgagtagtgggctcagtcggaacgta	100 nmole	HPLC
arm 6.3	gcttgtcggtagaacggtggaagcctccggtgtcgt	100 nmole	HPLC
arm 6.4	taacgctgttcgtggcgctatcattg	100 nmole	HPLC
arm 7.1	gtctgggtagcgtgggtaatcgacgt	100 nmole	HPLC
arm 7.2	cgttagctgaactgatgtggattcgcggttccaatg	100 nmole	HPLC
arm 7.3	aattcggtatggcgttgtgggatctgtcgattagac	100 nmole	HPLC
arm 7.4	aagctgatagagtggtacatcttgag	100 nmole	HPLC
arm 8.1	atgttggtagcgtgggtaatcaatgt	100 nmole	HPLC
arm 8.2	agcttgctgaactgatgtggattcgcggttcctcaa	100 nmole	HPLC
arm 8.3	gctgtggtatggcgttgtgggatctgtcgatgcagt	100 nmole	HPLC
arm 8.4	ttcatgatagagtggtacatcaatct	100 nmole	HPLC
arm 9.1	catgacggcgtgtggttgcatagcaa	100 nmole	HPLC
arm 9.2	aacgtacggatgagtagtgggctcagtcggactaac	100 nmole	HPLC
arm 9.3	tgctgtcggtagaacggtggaagcctccggttgcag	100 nmole	HPLC
arm 9.4	tcattctgttcgtggcgctattcaat	100 nmole	HPLC
arm 10.1	ctgtgcggcgtgtggttgcattgcac	100 nmole	HPLC
arm 10.2	atgctacggatgagtagtgggctcagtcggaatgac	100 nmole	HPLC
arm 10.3	tcagatcggtagaacggtggaagcctccggtgttga	100 nmole	HPLC
arm 10.4	acgttctgttcgtggcgctatgttag	100 nmole	HPLC
arm 11.1	gaattggtagcgtgggtaatcgtcta	100 nmole	HPLC
arm 11.2	cttacgctgaactgatgtggattcgcggttcttacg	100 nmole	HPLC
arm 11.3	agagcggtatggcgttgtgggatctgtcgatagctc	100 nmole	HPLC
arm 11.4	agcatgatagagtggtacatcgtcat	100 nmole	HPLC
arm 12.1	catggggtagcgtgggtaatccttgg	100 nmole	HPLC
arm 12.2	atactgctgaactgatgtggattcgcggttctatgt	100 nmole	HPLC
arm 12.3	aacatggtatggcgttgtgggatctgtcgatacatt	100 nmole	HPLC
arm 12.4	gtaaggatagagtggtacatccgtaa	100 nmole	HPLC
arm 13.1	cagcacggcgtgtggttgcatctgca	100 nmole	HPLC

Table A.2
Continued

Strand	Sequence	Scale	Purification
arm 13.2	ttagaacggatgagtagtgggctcagtcggattagt	100 nmole	HPLC
arm 13.3	aatagtcggtagaacggtggaagcctccggttagat	100 nmole	HPLC
arm 13.4	agtacctgttcgtggcgctattcaca	100 nmole	HPLC
arm 14.1	taactcggcgtgtggttgcattgtat	100 nmole	HPLC
arm 14.2	gtactacggatgagtagtgggctcagtcggatgtga	100 nmole	HPLC
arm 14.3	cacagtcggtagaacggtggaagcctccggtgtgca	100 nmole	HPLC
arm 14.4	tctagctgttcgtggcgctattagct	100 nmole	HPLC
arm 15.1	gctctggtagcgtgggtaatcgagct	100 nmole	HPLC
arm 15.2	ctagagctgaactgatgtggattcgcggttcagcta	100 nmole	HPLC
arm 15.3	attctggtatggcgttgtgggatctgtcgattgatt	100 nmole	HPLC
arm 15.4	actgtatagagtggtacatctgctt	100 nmole	HPLC
arm 16.1	ttatgggtagcgtgggtaatcaattg	100 nmole	HPLC
arm 16.2	acagtgctgaactgatgtggattcgcggttcaagca	100 nmole	HPLC
arm 16.3	ccatgggtatggcgttgtgggatctgtcgatccaag	100 nmole	HPLC
arm 16.4	taagagatagagtggtacatctaagt	100 nmole	HPLC

About the Authors

Chris Dwyer is an assistant professor of electrical and computer engineering and of computer science at Duke University. His research interests include self-assembling computer systems, nanoscale system design and simulation, nanoscale circuit design and CAD, self-assembling device fabrication, and DNA self-assembly. He received a B.S. in computer science and engineering at the Pennsylvania State University and an M.S. and a Ph.D. in computer science at the University of North Carolina–Chapel Hill. He is a member of the AAAS, ACM, ACS, and IEEE.

Alvin Lebeck is a professor of computer science and of electrical and computer engineering at Duke University. His research interests include high-performance microarchitectures, hardware and software techniques for improved memory hierarchy performance, multiprocessor systems, and energy-efficient computing. He received a B.S. in electrical and computer engineering, and an M.S. and a Ph.D. in computer science at the University of Wisconsin–Madison. He is a member of the ACM and IEEE.

Index

Accumulator, 13, 68, 94, 95, 98, 100, 110, 111, 113, 114, 122, 125, 129, 130, 149, 164, 165, 166, 167
Active network, 55, 61, 107, 111, 114, 130, 135, 163
Addition oracle. *See* Oracles
Addressing, 112, 113
Anchor, 108, 109, 116, 118, 119, 120, 121, 122, 128, 133–134, 135, 136, 140, 141, 142, 145, 148, 162, 173
Arbitration, 118, 176, 191
Arithmetic and logic unit (ALU), 82, 112, 123, 126, 131, 135, 136, 137, 138, 146, 147, 156, 157, 175, 176
ASIC, 56, 57, 59
Asynchronous, 108, 109, 118, 131, 135, 136, 138, 139, 140, 145, 146, 174, 178, 186
Assembler, 42, 44, 45
At-fabrication computation, 56, 57, 58, 59–60, 63–68, 70, 71, 73–76, 81
Atomic force microscope (AFM), 8, 11, 25, 26, 27, 28, 29, 132, 197, 198, 199

Back-off, 118
Base pair, 9, 14, 15, 20, 33, 35, 40, 56, 58, 173
Bidirectional, 108, 139
Bit-serial, 79, 81, 82, 102, 111, 114, 126
Bit-slice, 111, 112
Block-edit oracle. *See* Oracles
Bottleneck, 83, 120, 128
Bottom-up, 1, 2, 5, 7, 29, 51, 136, 164
Boundary nodes, 116
Broadcast, 115, 116, 117, 118, 121, 122, 124, 135, 137, 138, 139, 140–145, 153–157, 173–174, 177, 184
Built-in-self test (BIST), 176, 186

Carbon nanotube field effect transistor (CNFET), 35, 37, 38–39, 43, 47, 48, 52, 109, 122–123, 134, 139
Cavity, 8, 38, 39, 44, 45, 47, 48, 85, 132, 134
Cell, 35, 37, 38, 39, 84, 108, 109, 110, 116, 118, 119, 121, 122, 124, 126, 132, 136, 143, 144, 147, 149, 150, 152, 153, 154, 160, 163
Cell identifiers, 116
Chaining, 64, 110, 121, 122
Commissive defect. *See* Defects
Complementarity, 10
Complementary metal oxide semiconductor (CMOS), 5, 29, 35, 36, 37, 40, 47, 49, 51, 52, 89, 126, 129, 131, 135, 139, 152, 162, 171
Conditional execution, 113
Conditional store, 112, 113, 126
Configuration phase, 108, 129, 131, 135
Constant address, 112, 113
Content addressable memory (CAM), 63, 64
Contention, 110, 138, 157
Continuous-variable optimization, 92
Control logic, 88, 124, 137
Corrugation, 13, 14
Critical node logic, 175, 187
Custom layout, 43

Data encryption standard (DES), 102–103
Data parallel, 135, 143, 146, 153, 162, 163, 164, 187, 188, 194
Datapath, 138, 139
Deadlock, 114, 115, 118, 135
Deconfiguration, 141, 142
Defects
  commissive, 36, 37, 38

Defects (cont.)
  isolation of, 140–141, 172–174, 176, 179, 181, 182, 183–184
  models of, 117
  omissive, 36
  rates of, 13, 51, 53, 114, 116, 117, 119, 120, 124, 129, 131, 132, 141, 146, 160, 161, 163, 164, 172, 179, 183
  tolerance of, 3, 36, 37, 57, 110, 131, 132, 133, 135, 139, 160, 161, 163, 164, 172
Defect density. See Defects
Defect isolation. See Defects
Defect model. See Defects
Defect rate. See Defects
Defect tolerance. See Defects
Deoxyribonucleic acid (DNA)
  definition, 3, 7
  computing, 56, 58, 63, 82
  grids, 25, 37, 38–39, 132, 139, 202–203
  lattice, 11, 68, 132, 133
  origami, 23
  scaffolds, 33, 34–35, 39, 45, 65, 132
Depth-first traversal, 122, 140, 141, 174, 178
Design rules, 11–14
Distributed array multi-processor (DAMP), 58, 79, 80, 82, 83, 90–95, 98–103, 107, 163, 171
Double helix, 9
Double-stranded DNA, 9, 10

Energy*delay, 139
Execution packet, 110, 111, 118, 120–122, 125

Fail-stop, 141, 171–178, 181, 187
Failure modes, 5, 174, 180–185
Fault, 36, 38, 51, 110, 129, 146, 162, 178, 182
Fibonacci, 123–125, 127
Floorplan, 123, 137, 138, 139
Fluidic assembly, 83
Format, 111
Fragment
  of DNA, 28, 44, 69
  of code, 114, 121, 124, 126

Full-duplex, 89, 139, 147, 174
Functional unit, 110

Gaussian filter, 148, 150, 151
General purpose architecture, 107, 135, 152, 162
Generalized oracle. See Oracles
Generic linkers. See Linkers
Global memory, 108
Global optimization, 90, 92, 103
Gold colloid, 34
Gradient, 90–91, 115–124, 126–127, 140, 141, 173, 174, 177

Hamiltonian path, 56, 58, 69, 70
Hamiltonian path oracle. See Oracles
Hamming distance, 35, 38
Handshake, 135, 139, 146, 174, 178, 182
Hardware test, 176–179, 181, 186, 187
Head node, 141, 145
Header, 111, 115, 118, 119, 121, 125, 128
Helical twist, 9, 10, 25
Hierarchical assembly, 7, 8, 10, 11, 14, 16, 17, 23, 25, 29, 34, 39
Hybridization, 12, 33, 63, 202

IBM Blue Gene, 128
Immediate, 53, 95, 112
Implementation, 15, 33, 54, 55, 58, 59, 65, 82, 95, 102, 111, 115, 124, 126, 146, 152, 162, 192
Indirect address, 112–113
Indirect load, 114
Infinite backoff, 188, 191, 192, 193, 194, 195
Instruction buffer, 137–138, 144, 145, 146, 153, 157, 159, 175, 176
Instruction packet. See Packet
Instruction reuse, 154–155, 157, 160
Instruction set architecture (ISA), 54, 143
Interconnection, 39–40, 41, 55, 56, 79, 83, 84, 85, 88, 107, 108, 114, 129, 173
Interleave, 112, 121

Link, 108, 110, 114, 115, 117–118, 119, 121, 124, 126, 131, 132–133, 134, 135, 136, 138, 139, 140, 142, 147,

157, 162, 168, 171, 172, 173, 174, 177, 178, 187, 188, 189–195
Link arbitration. *See* Arbitration
Linkers
  generic, 25–26
  specific, 27
Livelock, 114, 115, 118
Load, 47, 64, 67, 81, 85, 88, 93, 96, 99, 112, 113, 114, 121, 124, 126, 128
Logarithmic accumulate, 166, 167
Logical network, 114
Logical ring, 131, 135, 136, 139, 162, 188
Lookup table, 56, 58, 59, 64
Loop unrolling, 126, 147, 167

Massive parallelism, 126
Matrix multiplication, 58, 147–159, 161, 164, 192, 194
Melting temperature, 9–10, 12, 13, 15
Memory packet, 110, 118, 119, 121
Memory port, 108, 109, 110, 119–121, 125, 127
Mesoscale assembly, 33
Metal oxide semiconductor field effect transistor (MOSFET). *See* CMOS
Metrics, 8, 11, 12, 14, 18, 23, 100
Microinstruction, 137, 143–146, 155, 165, 167, 176
Micro-scale, 5
Mixed-variable optimization, 92
Mode bit, 143
Motifs, 7, 8, 10–17, 21, 23, 25, 26, 27, 28, 29, 34

Nanoscale active network architecture (NANA), 55–56, 107, 111, 112, 114, 122, 124, 128–129, 135, 163, 171, 179, 187
Nearest-neighbor, 15, 29, 148, 162
NEC earth simulator, 91, 101, 103, 128
Network, 38, 40, 41, 55, 65, 80, 107, 108, 110, 111, 114–115, 117, 118, 119–120, 121, 122, 124, 125, 126, 127, 128, 129, 131, 132, 133–135, 138, 140–142, 143, 145, 146, 157, 160, 161, 162, 163, 164, 171, 172, 173, 175, 179, 183, 184–185, 186, 187–195

Node, 39–40, 49, 51–54, 56, 58, 69–70, 80, 89, 90, 92–94, 107–110, 124, 126, 127, 128, 129, 131–149, 152, 153, 154, 156, 157, 159–161, 162, 163, 164, 165, 171–195
Node controller, 80, 81, 82, 88, 93, 98–100
Nucleobases, 9
Nucleotide, 9, 11, 13, 14, 15, 23, 26

Omissive defect. *See* Defects
Opcode, 111, 126, 137, 138, 143, 144, 175
Operand stream, 112, 114, 118, 121, 122, 124–125
Oracles, 58, 63, 71, 77, 118
  addition, 64, 65, 66, 68, 70
  block edit, 70–71, 72, 73, 74, 75, 76, 77
  generalized, 13, 68, 115
  Hamiltonian path, 56, 58, 69–70

Packet, 107, 108, 110, 111, 112–113, 114, 115, 116, 117–122, 124–128, 129, 140, 141, 143, 145, 173, 179
Packet chaining. *See* Chaining
Parallel pattern search, 91
Parallelism, 8, 110, 122, 126, 146, 153, 163
Parasitics, 43
Partially functional nodes, 172, 174, 179, 187
Periodicy, 11, 13, 14
Permanent fault. *See* Defects
Photolithography, 2, 3, 4, 35, 40, 49
Physical network, 114, 163, 171
Pipelined, 111, 147, 148, 152, 162
Positional fault. *See* Defects
Postfabrication computation, 57, 58, 59, 60, 63, 72, 73, 74, 75, 76
Power density, 51, 131, 139, 146
Power-up circuit, 85, 86, 87
Predicated execution, 135
Predication, 135
Prefabrication computation, 56, 57
Processing element (PE), 79, 80, 81–82, 85, 88, 89, 97–100, 131, 135–136, 137, 140, 141–155, 157, 160–167, 188, 194
Programming, 114, 164, 188

Random constant, 81, 82, 98
Random network, 129, 131, 132, 133, 134, 135, 140, 142, 143, 160, 161, 164, 172, 173, 184, 185, 187
Reachable, 115, 117, 122, 126, 184, 185, 186, 191, 192, 193, 194, 195
Read only memory (ROM), 60
Red brick wall, 109, 129
Register file, 135, 136, 137, 138, 139, 143, 147, 156, 167, 175, 176
Regularity, 35, 37, 38
Repeat counter, 137, 143, 145
Reverse path forwarding (RPF), 115, 140, 161, 172, 173, 182
Ring gated field effect transistor (RG-FET), 95
Ringer circuit, 67, 80, 88
Route, 43, 119, 122, 132, 138–139, 173–174, 175, 176, 178

Self-assembled networks, 185, 187
Self-assembly, 1, 2, 3, 7, 9, 10, 11, 16, 28, 29, 33, 35, 36, 37, 40, 42, 45, 49, 51–52, 53, 56, 58, 60, 63, 64, 65, 73, 74, 76–77, 79, 80, 82, 83, 103, 107, 129, 131, 132, 140, 162, 163, 164, 171, 172, 173, 179, 182, 187, 188, 190, 191, 194, 195, 197, 199
Self-organizing defect tolerant SIMD architecture (SOSA), 55–56, 131, 135, 136, 140, 143, 145, 146, 147, 148, 149, 150, 151, 152–153, 158, 159–164, 171, 172, 173, 174, 175, 187, 188, 194
Sequencing instructions, 121–122
SEQUIN, 18
Shift, 59, 60, 81, 93, 112, 113, 114, 143, 144, 149, 152, 165, 186
Signature, 110
Simulation program with an integrated circuit emphasis (SPICE), 42, 43, 45
Single instruction multiple data (SIMD), 79, 80, 97, 131, 132, 135, 154, 159, 163, 164, 188
Single-stranded DNA, 9, 35, 200
Software test, 171, 172, 174, 177, 179, 181, 187

Spanning tree, 115, 117, 118, 119, 121, 122
Specific linkers. *See* Linkers
Specific-interaction energy measure (SEM), 12, 14, 15, 17–18, 19, 21–23
Stack, 85, 99, 112
Status bits, 81, 82, 113, 141
Sticky-end, 10, 11, 13, 14, 15, 16, 17, 18, 19, 20, 23, 24, 25–26, 27, 69, 203
Sticky-end sequences, 14, 15, 23, 24, 25, 27
Store, 54, 56, 73, 86, 94, 97, 112, 113, 114, 118, 121, 126, 128, 136, 139, 140, 141, 143, 145, 157, 199, 200, 202, 203
Strand reuse, 25
String matching, 123
Structure, 1, 2, 3, 4, 5, 7, 8, 9, 10, 11, 12, 13, 14, 15, 16, 18, 20, 23, 25, 26, 27, 28, 29, 33, 34, 35, 36, 37, 38, 39, 43, 45, 68, 80, 83, 84, 85, 86, 87, 88, 108, 114, 115, 118, 122, 129, 134, 140, 164, 191, 197, 198, 202, 203
Supramolecular, 25
Surface energy, 33
Synchronization microinstruction, 144, 145

Tail node, 141, 142, 145
Target-interaction likelihood measure (TLM), 12, 14–23
Temporal computing, 75, 77
Thermal ordering, 13–14
Thermodynamics, 2, 9, 11, 29, 45
Thiol, 34
Tiny encryption algorithm (TEA), 148
Top-down, 1, 2, 7, 33, 84, 132, 164
Topology, 124, 162, 163, 172, 185, 188, 191, 194
Topology generator, 191
Transceiver, 123, 124, 138, 139, 140, 157, 162, 172, 174–181, 185, 187, 188, 191, 192, 193, 194, 195
Transient fault, 51, 110, 162
Transistors, 9, 35, 37, 38, 45, 47, 80, 95, 131, 132, 133, 139, 173, 182

Index

Up*/down* routing, 115, 118, 121
Utilization, 122, 128, 136, 154, 161, 163, 188

Verboten sequences, 16
Via, 89, 108, 109, 115, 116, 117, 118, 119, 120, 122, 125, 126, 127, 133, 134, 140, 142
Virtual channel, 114, 118, 119, 121, 135, 138, 139, 174, 175, 177, 178

Virtual network, 115, 121
Voting mechanism, 110

Wormhole routing, 115

XTEA: Extended tiny encryption algorithm, 148, 152, 160

Related Artech House Titles

Design and Test of Digital Circuits by Quantum-Dot Cellular Automata, Fabrizio Lombardi and Jing Huang

Nanoelectronics Principles and Devices, Mircea Dragoman and Daniela Dragoman

Organic and Inorganic Nanostructures, Alexei Nabok

Semiconductor Nanostructures for Optoelectronic Applications, Todd Steiner, editor

Advances in Silicon Carbide Processing and Applications, Stephen E. Saddow and Anant Agarwal, editors

Nanotechnology Regulation and Policy Worldwide, Jeffrey H. Matsuura

Optics of Quantum Dots and Wires, Garnett W. Bryant and Glenn S. Solomon

Mathematical Handbook for Electrical Engineers, Sergey A. Leonov and Alexander I. Leonov

Electrical Engineering: A Pocket Reference, Heinz Schmidt-Walter and Ralf Kories

For further information on these and other Artech House titles, including previously considered out-of-print books now available through our In-Print-Forever® (IPF®) program, contact:

Artech House Publishers	Artech House Books
685 Canton Street	46 Gillingham Street
Norwood, MA 02062	London SW1V 1AH UK
Phone: 781-769-9750	Phone: +44 (0)20 7596 8750
Fax: 781-769-6334	Fax: +44 (0)20 7630 0166
e-mail: artech@artechhouse.com	e-mail: artech-uk@artechhouse.com

Find us on the World Wide Web at: www.artechhouse.com